P9-DNV-021

ORGANOMETALLIC SYNTHESES

Volume 2

Nontransition-Metal Compounds

Organometallic Syntheses

Edited by

JOHN J. EISCH
DEPARTMENT OF CHEMISTRY
THE STATE UNIVERSITY OF NEW YORK AT BINGHAMTON
BINGHAMTON, NEW YORK

R. BRUCE KING
DEPARTMENT OF CHEMISTRY
THE UNIVERSITY OF GEORGIA
ATHENS, GEORGIA

1. Transition-Metal Compounds, R. BRUCE KING
2. Nontransition-Metal Compounds, JOHN J. EISCH

ORGANOMETALLIC SYNTHESES

Volume 2

Nontransition-Metal Compounds

By

John J. Eisch

DEPARTMENT OF CHEMISTRY
THE STATE UNIVERSITY OF NEW YORK AT BINGHAMTON
BINGHAMTON, NEW YORK

1981

ACADEMIC PRESS
A Subsidiary of Harcourt Brace Jovanovich, Publishers

New York London Toronto Sydney San Francisco

ACADEMIC PRESS, INC.
111 Fifth Avenue, New York, New York 10003

United Kingdom Edition published by
ACADEMIC PRESS, INC. (LONDON) LTD.
24/28 Oval Road, London NW1 7DX

Library of Congress Cataloging in Publication Data

Eisch, John J.
Nontransition-metal compounds.

(Organometallic syntheses; v. 2)
Bibliography: p.
Includes index.
1. Organometallic compounds. I. Title.
II. Series.
QD411.E36 547'.05 80-26998
ISBN 0-12-234950-4 AACR2

PRINTED IN THE UNITED STATES OF AMERICA

81 82 83 84 9 8 7 6 5 4 3 2 1

For our daughter,
Margy Eisch
(1955–1976)
A pearl of great price

Contents

Foreword

Compounds containing carbon–metal bonds have come to occupy a prominent position in many phases of modern chemical research. During the past twenty-five years these organometallic compounds have served as key reagents in masterly synthetic achievements, both in organic chemistry and in inorganic chemistry. Moreover, because of their varied patterns of molecular structure, organometallic compounds are themselves the test subjects in many current physicochemical studies. These latter investigations continue to furnish an ever-deepening understanding of chemical bonding. Thus researchers of diverse backgrounds are faced with the problem of preparing pure organometallic compounds in a reliable fashion. Because of the specialized techniques involved in handling substances which are often poisonous, flammable, and sensitive to moisture and air, chemists engaging in such preparative work for the first time need guidance. Although several procedures for the more common organometallic compounds (e.g., Grignard reagents; alkyls of mercury, tin, and zinc; and ferrocene) have been included from time to time in the excellent compilations, "Organic Syntheses" and "Inorganic Syntheses," many desired procedures must be ferreted out of the voluminous research literature. Descriptions of essential techniques and apparatus, uncovered incidental to another research goal, often are difficult to trace by available literature indexes.

The editors have decided, accordingly, to prepare a work devoted to furnishing clear and reliable procedures for the preparation of important organometallic types. To this end, Volume 1 presented accurate descriptions for various transition-metal compounds, together with a general discussion of special laboratory techniques necessary in such experimentation (reactions under pressure or requiring ultraviolet irradiation; procedures for the purification, identification, and analysis of products; safety precautions; etc.). The second volume deals with nontransition-metal compounds, offering suitable procedures for this type, as well as a discussion of the purification of sensitive liquids, inert atmosphere provision, sample transfer, analytical techniques, etc.

The editors would like to acknowledge the fruitful discussions with the editorial board and with many of their research colleagues concerning the feasibility and format of this work. Finally, the editors and editorial board are gratified by the enthusiasm with which this endeavor has been greeted.

R. B. KING
J. J. EISCH

Preface

"Organometallic Syntheses" was conceived by the editors over fifteen years ago with feelings of great enthusiasm and resolve. In my esteemed coeditor, Bruce King, such resolve was matched by performance, and his volume on transition-metal organometallic synthesis was published in 1965. My own work on Volume 2 was beset by a swarm of unexpected professional distractions, which made completion of the manuscript a continually dimming prospect. That I can finally present Volume 2 is a matter of both satisfaction and relief. I only hope that the long delay has not exhausted the patience or the interest of the chemistry community.

The passage of time while a manuscript is in progress does have its merits: The Roman poet Horace advised that manuscripts could fittingly be stored in the author's desk for a decade or more, rather than rushed into publication. In my case, the delay has caused me to reexamine the original aims of the project in light of the appearance of several compendia and laboratory manuals dealing, in part, with organometallic preparations (see Appendix I). Such consideration has reconvinced me of the value of this work. In one place are offered reliable procedures that produce important organometallics or that can serve as models for preparing unknown organometallics. Furthermore, general experimental techniques for the convenient manipulation and characterization of such reactive substances are discussed or indicated in suitable detail for those unversed in such laboratory work. The fact that chemists from such diverse areas as organic, inorganic, analytical, structural, and biological chemistry have been attracted to the use of organometallic reagents shows how many scientists might find a work such as this highly useful.

In Part I of this volume we offer detailed directions and suggestions for (*a*) establishing and maintaining an inert atmosphere and solvent medium for the reaction and manipulation of organometallics; (*b*) evaluating the effects of purity, mode of mixing, solvent type, and conditions on the course of reaction; (*c*) monitoring, isolating, and purifying reaction products; (*d*) physical and chemical characterization of organometallic products; and (*e*) safe experimentation. Where extensive discussions would cover the same ground presented elsewhere, references to the original articles or reviews are simply given.

In Part II reliable procedures are offered for the synthesis of over 85 nontransition-metal organometallic compounds. With transition-metal organometallics, the nature of the organic ligand often largely determines the structure and reactivity of the compound. Consider the similarities among all cyclopentadienyl or all carbonyl derivatives. Main-group organometallics with the same organic ligands, on the other hand, often vary widely in their properties. In these cases, the nature of the metal predominates in determining the structure and reactivity. Therefore, it is deemed appropriate to classify these nontransition-metal organometallic syntheses by metal. For many metals the preparation of the alkyl, aryl, acetylenic, and allylic derivatives is described in detail. In addition, the preparations of various halo-, alkoxy, hydrido-, metallocyclic, and functionally substituted metal derivatives are included as examples of interesting metal–carbon systems. The author's research group has either developed these procedures or has had occasion to reproduce the procedure as originally given in the literature. In each instance, however, over the past twenty years, two or more of the author's research assistants have worked firsthand with these procedures. Hence the syntheses should be reliable and reproducible for a trained chemist.

With the broad commercial availability of many organometallics (see Appendix II), the question might naturally be asked, Why bother to synthesize? The generously funded research group requiring the occasional or routine use of a common organometallic might be well advised simply to purchase. But extended or more demanding needs for a given organometallic will often find commercial sources too costly, too slow, or too uncertain as to purity. Also, with increasing frequency chemists require isotopically labeled samples of organometallics in spectroscopic or mechanistic studies, and here the perusal of chemical catalogs will largely be fruitless or financially shocking. Adaptation of one of the procedures given here to isotopically labeled starting materials will then be most attractive.

Finally, the presentation of a detailed procedure for a commercially available organometallic is justifiable, in that it furnishes a typical procedure often readily adaptable to the synthesis of a rarer homolog or related structural type. The synthesis of triethylgallium from aluminum alkyls and gallium trichloride, for example, is easily extensible to the procuring of higher gallium alkyls, just as the synthesis of triphenylaluminum from diphenylmercury and aluminum is applicable to preparing a variety of aluminum aryls.

Securing a set of reliable organometallic synthetic procedures is a dynamic process, not a judgment valid for all time. Although the procedures described here and in Volume 1 should prove reproducible and serviceable, the coeditors would be indebted to any colleague who could suggest weak points or improvements in these procedures. As research and experience in organometallic synthesis broaden, surely many of the procedures given here will be modified or supplanted.

I wish to thank Professor Bruce King for his many helpful comments and his understanding patience with his dilatory colleague. Many of my research students have, by their diverse experience, detected pitfalls and enhanced the reliability of these procedures. Their individual contributions are noted throughout the text. Special gratitude is owed to Professor King and my co-workers, Dr. James E. Galle, Mr. Allen A. Aradi, and Mr. Lawrence E. Hallenbeck, who read the manuscript with a critical eye. Such an acknowledgment is not meant, however, to absolve me from ultimate responsibility for any remaining errors.

JOHN J. EISCH

Binghamton, New York
January 1980

PART I

General Experimental Techniques in the Organometallic Chemistry of Nontransition Metals

A. Establishment of Anhydrous and Anaerobic Reaction Media

1. General Reactivity Considerations for σ-Bonded Organometallics

a. Solvent or Reaction Medium

In reactions leading to the initial formation of a σ-carbon–metal bond[1] or in cases where one organometallic type is converted into a new organometallic system,[2] it is usually essential that oxygen and water be excluded. This caveat also applies to all recognized oxidants and to all potential proton donors, such as elemental sulfur, nitrogen oxides, disulfides, peroxides, halogens, and pseudohalogens (A—B), on the one hand, and acids, alcohols, partially alkylated amines, mercaptans, and terminal acetylenes (H—S), on the other. Furthermore, many inorganic and organic compounds containing multiple bonds (A=B: C≡C, C=O, C≡N, C=S, N=N, etc.) are susceptible to additions by organometallics (Scheme I).[1] Hence appropriate purification should be undertaken to remove all such adventitious reactants from the solvents and the atmosphere coming in contact with metal alkyls. Since metal alkyls and metal hydrides show great similarity in reactivity, the same purification procedures should be followed for metal hydrides as well.[3] In the most demanding preparations, for example with dialkylaluminum hydrides, all traces of the oxides of carbon, nitrogen, and sulfur, as

---

[1] See Appendix I for an extensive bibliography of the organometallic chemical literature. Chapters 2 and 5 of J. J. Eisch, "The Chemistry of Organometallic Compounds," Macmillan, New York, 1967, and Chapter I of G. E. Coates, M. L. H. Green, P. Powell, and K. Wade, "Principles of Organometallic Chemistry," Methuen, London, 1968, are concise, useful discussions of carbon–metal bond reactivity.

[2] (a) R. G. Jones and H. Gilman, *Chem. Revs.* **54,** 835 (1954); (b) J. J. Eisch, "The Chemistry of Organometallic Compounds," Chapter 6, Macmillan, New York, 1967; (c) G. E. Coates, M. L. H. Green, P. Powell, and K. Wade, "Principles of Organometallic Chemistry," Chapter 2, Methuen, London, 1968.

[3] Not only can impure solvents spoil a metal alkyl or metal hydride reaction, but also an explosion or fire can result. Unsaturated hydrocarbons and anhydrous ethers rapidly form hydroperoxides and peroxides whose reaction with metal hydrides or with certain metal alkyls can be violent (Section A,2).

well as carbonyl, olefinic, and acetylenic impurities, should be rigorously excluded.

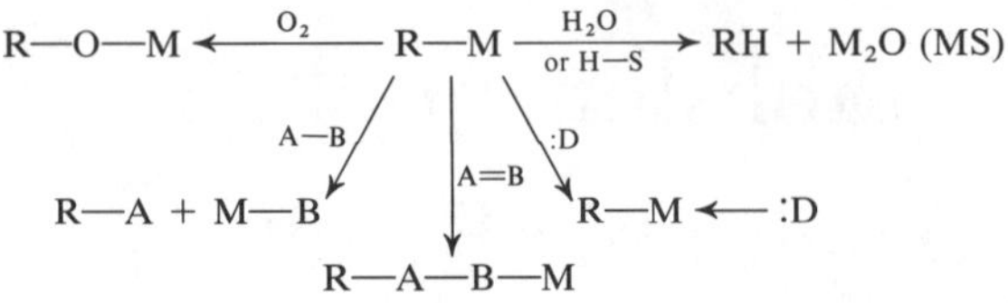

$$\mathrm{R{-}O{-}M} \xleftarrow{O_2} \mathrm{R{-}M} \xrightarrow[\text{or H—S}]{H_2O} \mathrm{RH + M_2O\ (MS)}$$

$$\mathrm{R{-}M} \xrightarrow{A-B} \mathrm{R{-}A + M{-}B}$$

$$\mathrm{R{-}M} \xrightarrow{A=B} \mathrm{R{-}A{-}B{-}M}$$

$$\mathrm{R{-}M} \xrightarrow{:D} \mathrm{R{-}M \leftarrow :D}$$

SCHEME I

Finally, in striving for a completely inert medium one should bear in mind that either the solvent or a metal salt by-product may react by complexation with the desired metal alkyl product. Donor solvents, such as ethers, sulfides, and amines, and even other metal alkyls can form stable stoichiometric solvates which exhibit properties distinctly different from those of unsolvated metal alkyls and which often lose their solvent molecules only under conditions that destroy the metal alkyl. Illustrative is the preparation of tri-*tert*-butylaluminum from the Grignard reagent and aluminum(III) chloride in ethyl ether solution. An attempt to remove the ether by heating the distillable aluminum alkyl etherate at higher temperatures leads to the decomposition.[4]

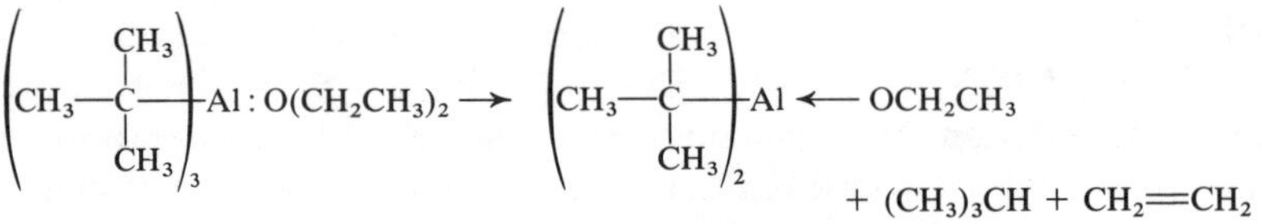

$$\left(\mathrm{CH_3{-}\overset{CH_3}{\underset{CH_3}{C}}}\right)_3\mathrm{Al:O(CH_2CH_3)_2} \longrightarrow \left(\mathrm{CH_3{-}\overset{CH_3}{\underset{CH_3}{C}}}\right)_2\mathrm{Al \leftarrow OCH_2CH_3} + \mathrm{(CH_3)_3CH + CH_2{=}CH_2} \qquad (1)$$

In many other instances, the solvent used for preparation of the organometallic may be unsuitable for its storage. The limited stability of ethers to cleavage by organolithium reagents must be borne in mind.[5,6] Especially noteworthy is the cleavage of tetrahydrofuran (THF) by lithium alkyls into ethylene and lithium ethenoxide. The latter fragments can cause the formation of unusual side products[7,8]:

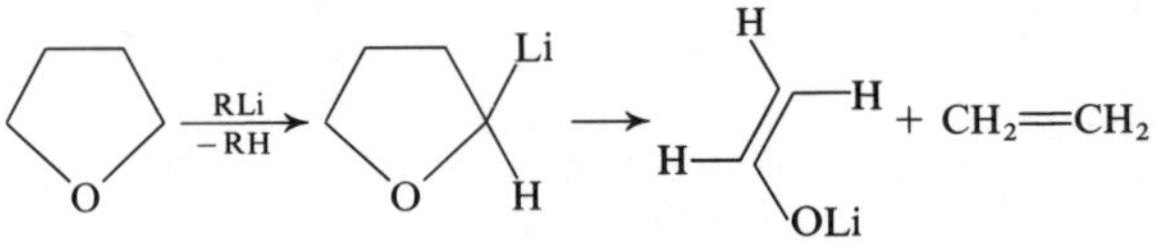

$$\text{THF} \xrightarrow[-RH]{RLi} \text{2-lithio-THF (C(Li)(H))} \longrightarrow \mathrm{H_2C{=}C(H)OLi} + \mathrm{CH_2{=}CH_2} \qquad (2)$$

$$\text{THF} \xrightarrow{n\text{-BuLi}} \xrightarrow{Ph_2C=O} \mathrm{Ph_2\overset{OLi}{C}{-}CH_2{-}C{\overset{O}{\underset{H}{\lessgtr}}}} \qquad (3)$$

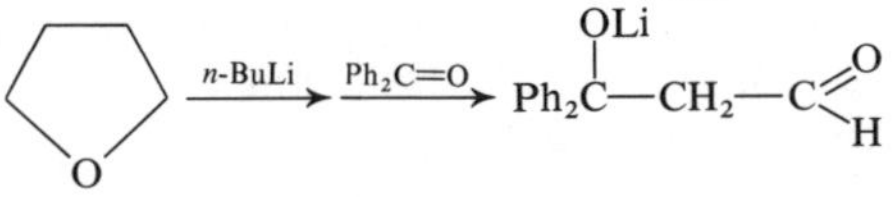

---

[4] H. Lehmkuhl, *Justus Liebigs Ann. Chem.* **719,** 40 (1968).

[5] H. Gilman, A. H. Haubein, and H. Hartzfeld, *J. Org. Chem.* **19,** 1034 (1954).

[6] U. Schöllkopf, *in* "Houben-Weyl Methoden der Organischen Chemie" (E. Müller, ed.), Vol. XIII/1, p. 19. Georg Thieme, Stuttgart, West Germany, 1970.

In a related fashion, metal salts present in organometallic systems can form aggregated complexes of the type $[R_mM]_a \cdot [MX_m]_b$, which will have physical and chemical properties quite different from those of R_mM. Comparison of Grignard reagents with R_2Mg,[9] lithium halide-containing lithium alkyls with pure lithium alkyls,[10] and aluminum alkyls with their alkali metal halide complexes $(MX \cdot R_3Al)$[11] shows how important such complexation can be. These complexes can either hold the metal salt tenaciously or undergo a redistribution reaction [Eq. (4)],[12] thereby hindering isolation of the halide-free alkyl:

$$R_3Al \cdot AlCl_3 \rightleftharpoons RAlCl_2 + R_2AlCl \tag{4}$$

Just as organometallics can form stable complexes less reactive than the unsolvated metal alkyl, the synergistic effect of dissimilar metal alkyl pairs (e.g., tritylsodium and triphenylborane on THF[13] [Eq. (5)]) serves to warn the experimenter that the establishment of an "inert" medium is often a subtle guessing game involving factors such as the reactivity and reaction mechanism of a particular metal compound, the time of contact between the candidate medium and the reagent, the reaction temperature, and the presence of potential coreactants.

$$\text{N.R.} \xleftarrow{(C_6H_5)_3CNa} \text{THF} \xrightarrow[(C_6H_5)_3B]{(C_6H_5)_3CNa} (C_6H_5)_3C{-}CH_2{-}CH_2{-}CH_2{-}CH_2{-}O^-Na^+ \tag{5}$$

The most reactive metal alkyls (Group IA) require a saturated hydrocarbon as a solvent or, for truly saline alkyls, as a dispersing agent. By using as the hydrocarbon solvent the Brønsted acid of the organometallic, solvent attack can be accommodated. Thus phenylsodium suspended in benzene can readily bear transmetalation with the medium. (The disadvantage of encountering certain amounts of di- or polymetalated derivatives of the solvent, however, may have to be borne.)

b. Reaction Atmosphere

For an inert cover for the majority of organometallic reactions and manipulations, purified, dry nitrogen is the most convenient and economical.

---

[7] R. B. Bates, L. M. Kroposki, and D. E. Potter, *J. Org. Chem.* **38,** 322 (1973).

[8] P. Tomboulian *et al.*, *J. Org. Chem.* **38,** 322 (1973).

[9] (a) R. M. Salinger, *in* "Survey of Progress in Chemistry" (A. F. Scott, ed.), Vol. 1, p. 301, Academic Press, New York, 1963; (b) E. C. Ashby, J. Laemmle, and H. M. Neumann, *Acc. Chem. Res.* **7,** 272 (1974).

[10] (a) W. Glaze and R. West, *J. Am. Chem. Soc.* **82,** 4437 (1960); (b) T. L. Brown, *Adv. Organometal. Chem.* **3,** 365 (1965).

[11] K. Ziegler, R. Köster, H. Lehmkuhl, and K. Reinert, *Justus Liebigs Ann. Chem.* **629,** 33 (1960).

[12] H. Lehmkuhl and K. Ziegler, *in* "Houben-Weyl Methoden der Organischen Chemie (E. Müller, ed.), Vol. XIII/4, p. 53, Georg Thieme, Stuttgart, West Germany, 1970.

[13] G. Wittig and O. Bub, *Justus Liebigs Ann. Chem.* **566,** 113 (1950).

For less reactive organometallics, such as zinc alkyls, carbon dioxide or even hydrogen can be used. The greater density of carbon dioxide as compared to that of hydrogen makes the former more suitable for displacing air from vessels and for surrounding a sensitive substance with an inert layer upon transfer. For the same reason, diethyl ether often can serve as both solvent and inert cover for metal alkyls made in solution for immediate use (for example, Grignard reagents prepared in air).

The most reactive organometallics, such as alkali metal alkyls and aluminum alkyls, cannot be exposed to ethers, hydrogen, carbon dioxide, or even small amounts of oxygen. Either a violent reaction will occur or, at a minimum, the alkyl will be destroyed. Even gaseous nitrogen is not a suitable atmosphere for the most demanding cases. Lithium metal at room temperature and both magnesium and calcium metals at elevated temperatures combine with nitrogen to form nitrides; such a reaction can be catalyzed by transition-metal impurities from stirrers or reaction vessels.[14] This nitridation can retard desired reactions by coating the metal surface. Furthermore, a wide variety of transition-metal organometallics form complexes with molecular nitrogen, and in certain cases can even reduce nitrogen to nitrides or hydrazine salts.[15]

Recourse to either helium or argon therefore is advisable for protecting the most sensitive organometallics. The greater density [sp. gr. 1.38 (A)] or "pourability" of argon makes it the inert cover of choice.

c. Empirical Gradations in Reactivity

Extensive studies have been conducted that compare the proton-abstracting strength, the oxidizability, the tendency to add to carbonyl and related functions, the thermal instability, and the Lewis acidity of main-group organometallics. The electronic reasons behind these empirical trends are complex and incompletely understood. However, for the experimentalist certain broad generalizations about reactivity should be helpful in gauging the difficulty and sensitivity of the laboratory task.

The reactivity rules formulated by Gilman,[16] on the basis of additions to carbonyl substrates, can generally be understood in terms of the polarity of the carbon–metal bond. The more polar the bond, or the less electronegative the metal, the more reactive the metal alkyl.[17] Accordingly, alkyls

[14] K. Shiina, *J. Am. Chem. Soc.* **94,** 9266 (1972).

[15] M. E. Vol'pin and V. B. Shur, *in* "Organometallic Reactions" (E. I. Becker and M. Tsutsui, eds.), Vol. 1, p. 55, Wiley (Interscience), New York, 1970.

[16] H. Gilman, *in* H. Gilman (ed.), "Organic Chemistry—An Advanced Treatise," Vol. 1, 2nd ed., p. 520. Wiley, New York, 1943.

[17] E. G. Rochow, D. J. Hurd, and R. N. Lewis, "The Chemistry of Organometallic Compounds," p. 21, Wiley, New York, 1957.

of the alkali metals are the most sensitive to oxidation and to cleavage by protic sources. Within Group IA, reactivity increases in the expected order: Li < Na < K < Rb < Cs. The polar character and reactivity of the group are such that most alkyls are insoluble in hydrocarbons and readily attack donor solvents such as ethers and amines.

Conversely, alkyls formed from metals of greater electronegativity have sharply reduced reactivity. Although still sensitive to air oxidation, boron alkyls react only slowly with water, and the alkyls of Group IVA, Group VA, and mercury generally have carbon–element bonds stable to moisture and to oxygen.

Generally, organometallics of Groups IA, IIA, and IIIA react readily with oxygen, with water, and with Lewis bases. With the aforementioned exception of mercury, the same reaction patterns hold for the alkyls of Groups IB and IIB. Although Group IVA alkyls are stable to water and Lewis bases, their halide, alkoxide, and hydride derivatives are not. The alkyls of Group VA can form oxides of the type $R_3E \rightarrow O$ in air, and their halide (R_nEX_{3-n}) and hydride derivatives (R_nEH_{3-n}) are also prone to hydrolysis.

For organometallics of a given metal, the reactivity increases in the following sequences: (*a*) for alkyl derivatives, *n*-alkyl < *sec*-alkyl < *tert*-alkyl; (*b*) for varying hybridizations of carbon in the carbon–metal bond, $sp < sp^2 < sp^3$; and (*c*) for alkenyl groups, other alkenyls < vinylic < allylic, or for aryl-containing groups, *β*-arylalkyl or further removed < aryl < α-arylalkyl.

2. Inert Gases

a. Purification

The most commonly used inert gases in organometallic chemistry, in order of decreasing utility, are argon, helium, nitrogen, hydrogen,[18] and carbon dioxide,[19] with the limitation that the last-mentioned two may be used only with the least reactive metal alkyls. These gases can be purchased in high purity and may be used directly for less critical situations, such as

[18] The phenyl derivatives of lithium, sodium, potassium, rubidium, cesium, and calcium undergo hydrogenolysis when treated with 1–2 atm of hydrogen at room temperature [H. Gilman, A. L. Jacoby, and H. Ludeman, *J. Am. Chem. Soc.* **60,** 2336 (1938)].

[19] The alkali metal and alkaline earth organometallics react extremely vigorously with carbon dioxide to form acids and ketones; organometallics of Group IIIA sometimes require heat and pressure for the carbonation of even one carbon–metal bond. In fact, attempted carbonation of the diisobutyl(*β-cis-* and *trans*-stilbenyl)aluminums has revealed an interesting case of steric hindrance to carbonation of the *β-trans*-stilbenylaluminum bond [J. J. Eisch and M. W. Foxton, *J. Organometal. Chem.* **11,** P7 (1968)].

the preparation of Grignard or organolithium reagents in solution. Indeed, it may be generally better to use highly prepurified gases directly from the cylinder than to subject the gas stream to a poorly devised purification train. So that the emerging gas from a purification apparatus will not be *less pure* than the original gas, considerable care in building the purification unit is required. Ordinary rubber tubing, which is porous to oxygen and moisture, must be avoided; provision for possible aerosol transport of the liquid and solid purifying agents must be made; and the limitations that the gas flow rate, the rates of deoxygenation and dehydration exhibited by purifying agents, and the capacity of various absorbents for impurity removal must not be ignored. Periodic mass spectrometric analysis of the "purified" gas will demonstrate whether the purification process continues to be effective.

Choice of a purification method for the inert gas, among the many that have been described, should be governed by the following factors: (*a*) the effectiveness of the deoxygenant and the desiccant at moderate rates of gas flow; (*b*) the irreversible (chemical) removal of oxygen and water to a minimum ($<0.0001\%$) level; (*c*) the assurance that no new impurities (e.g., dust, CO, H_2, hydrocarbons) will be introduced in the process; (*d*) the high capacity of a given absorbent for removing impurities; and (*e*) the ease of regenerating the purification train. The last-mentioned two points bear on the convenience of a method, the former three on the essentials. In most procedures, deoxygenation is done in one step involving contact with water (aqueous solutions) or the formation of water (CuO in the presence of H_2). Hence, in a subsequent step, the gas stream must be dried. However, the use of neat, high-boiling metal alkyls[20] (e.g., the adduct of acetylene and triethylaluminum or $[(C_2H_5)_2AlCl]_2$ by itself) permits both dehydration and deoxygenation to be achieved simultaneously. Naturally one must be certain that the resulting volatile hydrocarbon (RH) will not interfere with the emerging inert gas. Passage of the gas through a solid carbon dioxide cold trap at $-78°$ usually removes most of these impurities.

Although several deoxygenating solutions, such as pyrogallol,[21,23] anthraquinone β-sulfonate,[22] hydroxyhydroquinone,[23] sodium dithionite,[24] chromium(II) salts,[25] titanium(III) chloride,[26] vanadium(IV) sulfate,[27]

---

[20] See K. Ziegler and H.-G. Gellert, U.S. Patent 2,695,327, November 23, 1954 [*Chem. Abstr.* **50,** 1073 (1956)].
[21] F. Henrich, *Ber.* **48,** 2006 (1915).
[22] L. F. Fieser, *J. Am. Chem. Soc.* **46,** 2639 (1924).
[23] F. Henrich and K. Kuhn, *Z. Angew. Chem.* **29,** 150 (1916).
[24] J. Thiele, *Ber.* **31,** 1247 (1898).
[25] E. E. Aynsley and P. L. Robinson, *J. Chem. Soc.* 58 (1935).
[26] L. Moser, *Z. Anorg. Chem.* **110,** 130 (1920).
[27] L. Neites and T. Neites, *Anal. Chem.* **20,** 984 (1948).

and copper(I) salts,[28] have been recommended, these reagents have convenience, rather than maximum effectiveness, as their chief virtue. For occasional work under a nitrogen atmosphere, employing ordinary cylinder gas (99.7%), one might resort to such reagents, but for extensive work with sensitive metal alkyls (R_3B), one is advised to employ heated or activated copper metal. Such a deoxygenating agent best fulfills the aforementioned criteria. Typically, a bed of finely divided copper (small chips or gauze) is maintained at 350–400°[29] or, alternatively, a freshly reduced charge of finely divided copper on an inert carrier (BTS catalyst[30]) functions at room temperature in reducing the oxygen content of the nitrogen gas to a very low level. The color of the reduced copper (shiny for copper; coal black for the BTS catalyst) serves to indicate the continuing capacity of the charge; and periodic treatment with hydrogen suffices to regenerate the charge (see below).

The removal of moisture from deoxygenated gas can best be achieved by passage through a porous column of phosphorus pentoxide. Care must be taken to prevent the syrupy phosphoric acid formed from coating the surface completely and decreasing the dehydrating activity. Dispersion of phosphorus pentoxide on glass beads or the use of granular, porous pellets[31] is most effective.

b. Experimental Setup for the Purification and Distribution of an Inert Gas

The gas purification and distribution train shown in Figs. 1–3 has been used successfully, with various modifications, in several organometallic research laboratories. The apparatus consists of a mineral oil bubbler (A) and a mercury excess-pressure valve (C) (Fig. 1); a deoxygenating and drying train (Fig. 2); and a distribution manifold and a mercury manometer (Fig. 3). Cylinder gas is passed (B_2) through a heated glass tube (or unheated, if the BTS catalyst is used) containing the copper metal (D in Fig. 2) and then through a tube containing anhydrous calcium sulfate[32] or a synthetic metal aluminosilicate[33] for preliminary drying (E). For final purification the gas is led through a tube containing a porous bed of phosphorus pentoxide (F)

---

[28] H. von Wartenberg, *Z. Elektr. Chem.* **36,** 295 (1930).

[29] F. Hein, E. Petzschner, K. Wagler, and F. A. Segitz, *Z. Anorg. Chem.* **141,** 190 (1924).

[30] Available from BASF Colors and Chemicals, Inc., 866 Third Avenue, New York, New York 10022.

[31] Available as granular phosphorus pentoxide of 99.5% purity under the trademark Granusic from J. T. Baker Chemical Company, Phillipsburg, New Jersey 08865.

[32] Available in various mesh sizes as Drierite from the W. A. Hammond Drierite Co., 138 Dayton Ave., Xenia, Ohio 45385.

[33] Available in pellet or rod shapes as Linde molecular sieves from Union Carbide Corp., Linde Division, 270 Park Ave., New York, New York 10017.

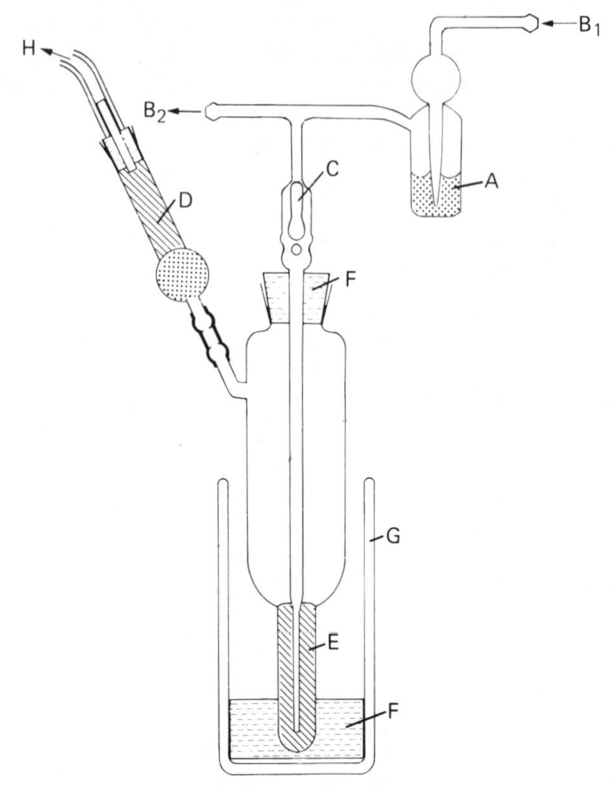

FIG. 1. Gas bubbler and excess-pressure valve. A, Mineral oil bubbler; B_1, inert gas from cylinder; B_2, exit gas; C, floater valve with ground-glass upper surface; D, mixture of sulfur and glass wool; E, mercury; F, cork support; G, metal receptacle; H, tube exiting to hood.

and, if desired, a final cold trap of solid carbon dioxide–ethanol (G) to remove any volatile organic matter. At A in Fig. 3 a vacuum can be connected to the distribution manifold and thereby to a reaction vessel; a standpipe manometer (C) indicates positive or negative pressure in the system. The bubbler (A in Fig. 1) should have a bulb reservoir to prevent oil from backing up into the gas cylinder exit tube. Mineral oil for such bubblers should be heated with molten sodium and allowed to cool over the metal under a dry nitrogen atmosphere. The heated Pyrex glass tube (D in Fig. 2; 25 cm o.d., 180 cm long) should be positioned horizontally in an ordinary mantle-type combustion furnace whose aperture permits about one-third of the copper to be heated at one time. As the copper charge becomes oxidized through use, the furnace is positioned over an unoxidized section of D. Activation of such a copper metal-packed column is achieved as follows. The tube is packed vertically with copper gauze or light turnings (previously washed with benzene and dried thoroughly), the metal being held in place

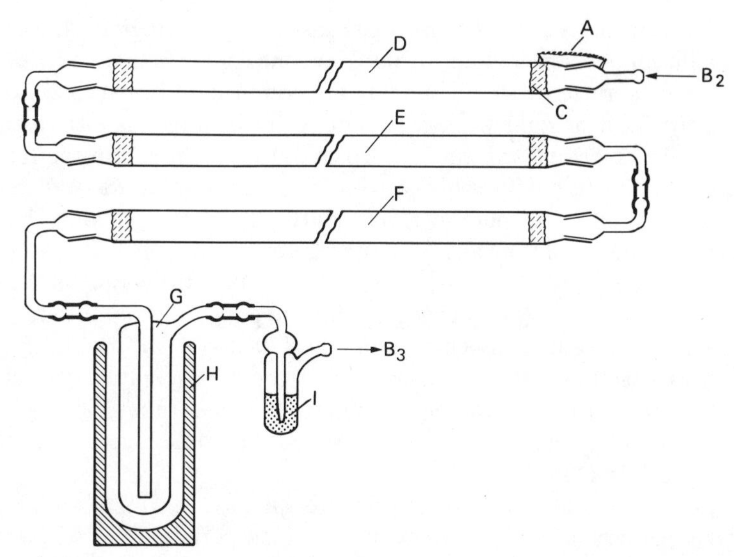

FIG. 2. Purification train for an inert gas. A, Steel springs (on all joints); B_2, entering inert gas; B_3, exiting inert gas; C, glass wool end packing; D, copper or BTS catalyst; E, anhydrous $CaSO_4$ or molecular sieves; F, powdered P_2O_5; G, cold trap; H, Dewar flask; and I, bubbler.

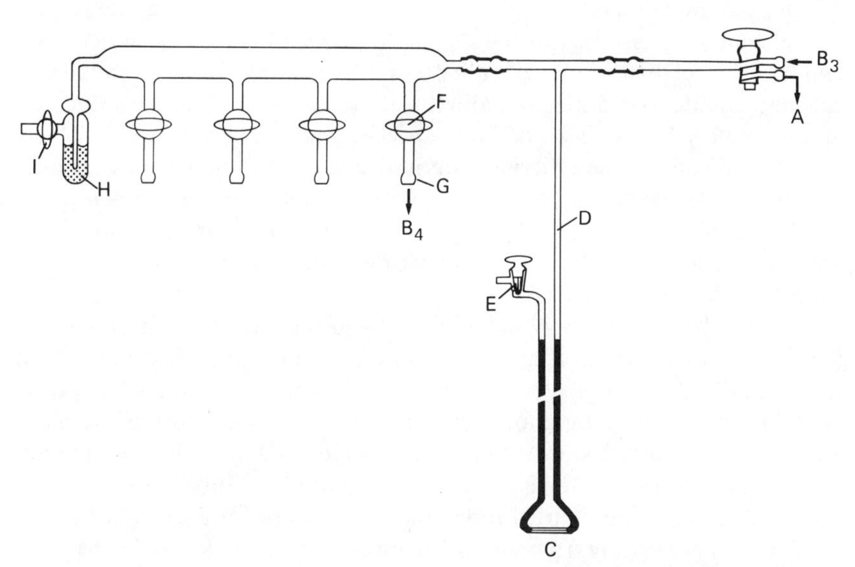

FIG. 3. Inert gas distribution manifold and manometer. A, Vacuum pump; B_3, entering inert gas; B_4, exiting inert gas; C, manometer with capillary restriction; D, manometer (scale); E, valve for closure when the manifold is evacuated; F, two-way stopcock; G, knurled tube for rubber tube connection; H, mineral oil bubbler; I, stopcock.

with copper or steel gauze; then nitrogen gas is swept through the horizontal tube as the copper is heated up to 200–300°; the copper is then reduced to a bright, shiny surface by slowly moving the combustion furnace, section by section, from left to right while a stream of dry hydrogen is introduced at the left end; the condensed water is driven out the right end of the tube by supplemental heating with a hot-air blower as the tube is allowed to cool under an atmosphere of pure nitrogen. Naturally, safety demands that the exit gas be removed by means of a rubber tube to an efficient hood. The tower (D) is now ready for connection to the train. If the copper is the BTS catalyst, then the tube (D) is charged with the gray pellets and reduced as described above, but at a temperature of 120–140°. The resulting coal-black deoxygenating agent will function at room temperature; hence the heating mantle for tube D can be omitted. In this case, tube D in Fig. 2 can be placed vertically. Any disposal of the reduced pyrophoric catalyst should be preceded by wetting.

In place of the mercury excess-pressure valve (C in Fig. 1) an 800-mm capillary mercury standpipe can be used, which permits reduced pressure in the system without ingress of air. For safety, the exit tube of C (or the mercury standpipe) can be connected to a drying tube (D), which is filled with a mixture of sulfur powder and glass wool. This precaution minimizes the escape of mercury vapor into the laboratory.

Drying towers E and F (Fig. 2) (2.5–4.0 cm o.d., 180 cm long) are packed with the aid of glass plugs at their ends, to prevent the settling of desiccant and the spray of dust. Tower E is filled with a 5-cm layer of 8-mesh anhydrous calcium sulfate containing a cobalt salt (Indicating Blue Drierite), an approximately 150-cm layer of white Drierite, and finally with another 5-cm layer of Indicating Blue Drierite. Alternatively, the 150-cm layer can consist of Linde molecular sieves, type 4A. Tower F is filled with a slurry of powdered phosphorus pentoxide and glass beads (4 mm diam.), or the use of granular pellets of phosphorus pentoxide[31] may be advantageous as a substitute.

These tubes are equipped with outer 24/40 standard-taper ground-glass joints for ease in disconnecting. The gas inlet and outlet joints fit 8-mm rubber tubing and are held tight to the tubes by silicone grease and steel springs (Fig. 2). The rubber tubing for all connections is kept as short as practical, and only butyl rubber stock (8 mm i.d.) is employed, in order to minimize atmosphere diffusion into the inert gas stream. Glass tube connections to rubber always require knurled tubes and wire clamps for a gastight fit.

The inert gas exiting from the purification train is then led to the distribution manifold (Fig. 3) which provides several gas outlets (G), a manometer (C), an excess-pressure valve (H), and a stopcock (A) for evacuation of the manifold and any of its connecting vessels. Each gas outlet (G) consists of an 8-mm-bore two-way stopcock bearing a knurled glass connection (8 cm long, 8 cm bore) for a length of butyl rubber tubing. So that C may function as a reduced-pressure gauge, the capillary tube (1 mm i.d.) (O) should be $\sim$80 cm high. Finally, the stopcock at A should be a three-way type of oblique bore, so that evacuation of the manifold and the readmission of inert gas can be performed alternately. For evacuation through G the stopcocks at A and I are shut off to isolate the manifold. The stopcock at A is then turned to make inert gas available at B_3. Thus, by turning stopcock I (manifold closed) and putting the stopcock at A alternately on vacuum and on line B_3, the manifold and its connectives can be flushed with inert gas. Finally, the ends of the butyl rubber connections from the manifold should be sealed by glass rodding when not in use.

c. Maintenance of an Inert Atmosphere

Although particular procedures for maintaining an inert atmosphere will be dealt with under the isolation and transfer of organometallics, several general points can be made. First, once a closed vessel is connected to the gas manifold, it can be evacuated and filled with inert gas *ad libidum*. In the evacuation phase it is useful to warm the reaction apparatus gradually and thoroughly with a luminous flame or a hot-air blower (hair dryer variety). Thereafter the reaction vessel need only be filled with nitrogen and kept at atmospheric pressure via valve F. A slight overpressure in the gastight system is ensured by bubbling at the manifold exit (I). This type of connection is most general. As an alternative, the butyl rubber connection can be used as an inlet to a reaction vessel, while the mercury or mineral oil bubbler shown in Fig. 4 can serve as both a gas inlet and outlet once an inert atmosphere is established. Thus, once a vessel is flushed as above, this bubbler can be surmounted on the condenser outlet and serve to admit gas or release excess pressure.

d. Inert Atmosphere Chamber ("Dry Box")

Although organometallic researchers have not completely realized A. von Grosse's suggestion that they fill their laboratories with nitrogen and don diving suits,[34] the development of various types of inert atmosphere

[34] E. Krause and A. von Grosse, "Die Chemie der Metallorganischen Verbindungen," p. 802, Verlag Borntraeger, Berlin, 1937. Within the last few years a metallurgical firm in Bremen, Federal Republic of Germany, has developed a process for coating metals with titanium, tantalum, zirconium, nickel, or other metals, which involves the workmen donning astronaut-type suits and conducting their coating procedure in a chamber filled with argon.

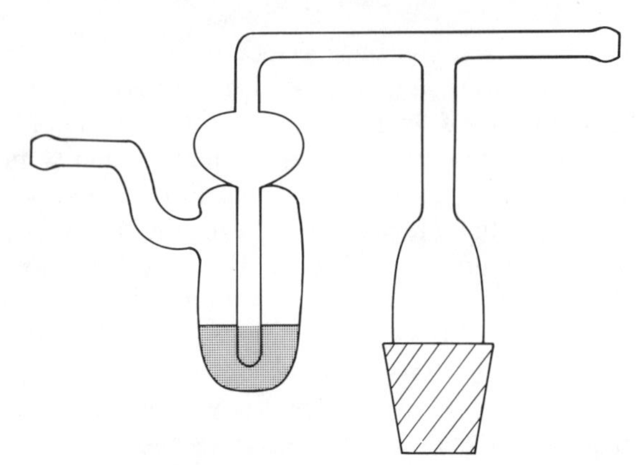

FIG. 4. Gas bubbler closure for pressure release.

glove boxes is a clear offspring of such a notion. In cost and effectiveness, inert atmosphere chambers range from cheap, disposable plastic bags, which can envelop equipment for a given transfer or measurement and can be evacuated and refilled with inert gas,[35] to expensive and elaborate metal chambers, which recirculate the atmosphere continuously through a purification train.[36] Among these and other commercially available models,[37] one could advise: "costly thy habit as thy purse can buy," but one could also add "costly as thy experiment requires." Quantitative studies justify more extensive expenditures than preparative investigations. Studies on electrochemical,[38] electron spin resonance,[39] and nuclear magnetic resonance (nmr) properties[40] of organometallics have employed the Dri-Lab inert chamber,[36] either with a standard purification and recirculation train (Dri-Train) utilizing a heated copper and molecular sieve system or with a special manganese(II) oxide–molecular sieve train. The details of the latter

---

[35] Polyethylene glove bags are available in various sizes from the Alfa Division of Ventron Corp., Danvers, Massachusetts 01923.

[36] Vacuum Atmospheres Co., Hawthorne, California 90250; Kewanee Scientific Equipment Co., Adrian, Michigan 49221.

[37] Contamination Control, Inc.; Coy Laboratory Products; Forma Scientific, Inc.; A. Gallenkamp and Co., Ltd.; Hamilton Industries; HEMCO Corp.; High Vacuum Equipment Corp.; Hotpack Corp.; Instruments for Research and Industry; Kahl Scientific Instrument Corp.; Labconco Corp.; Lab Glass, Inc.; Manostat Corp.; Nuclear Equipment Chemical Corp.; Perkin-Elmer Corp.; Raytheon Co. Industrial Electronics Dept.; Shimadzu Seisakusho, Ltd.; The Torsion Balance Co.

[38] R. E. Dessy *et al.*, *J. Am. Chem. Soc.* **83,** 453, 460, 467, 471 (1966).

[39] R. E. Dessy and R. L. Pohl, *J. Am. Chem. Soc.* **90,** 1995 (1968).

[40] G. E. Parris and E. C. Ashby, *J. Am. Chem. Soc.* **93,** 1206 (1971).

purification train, as developed by T. L. Brown and associates at the University of Illinois,[41] are given below.

The discussion of the inert atmosphere system will be based on the lettering of various components as shown in Figs. 5–8.

The Enclosure. The system requires a glove box (Fig. 5) which may be nonevacuable but should have an evacuable antechamber. The major part of the chamber may well be evacuable; it is, however, important that the antechamber be so, a feature that is regrettably absent from many enclosures presently offered on the market. The Model HE-43-2 Dri Lab manufactured by Vacuum/Atmospheres Corp. has been found to be eminently satisfactory.[36]

Purge Atmosphere. Most commonly argon has been used as a purge gas, but a good grade of nitrogen has also been employed with good results. Helium is not recommended for general use; it is lost very rapidly by diffusion through the gloves and is quite expensive.

Scavenging Pump and Enclosure. The box atmosphere is continuously circulated through a series of columns, to be described below, by a small oil-less reciprocating pump. Because such pumps (and in fact all pumps that

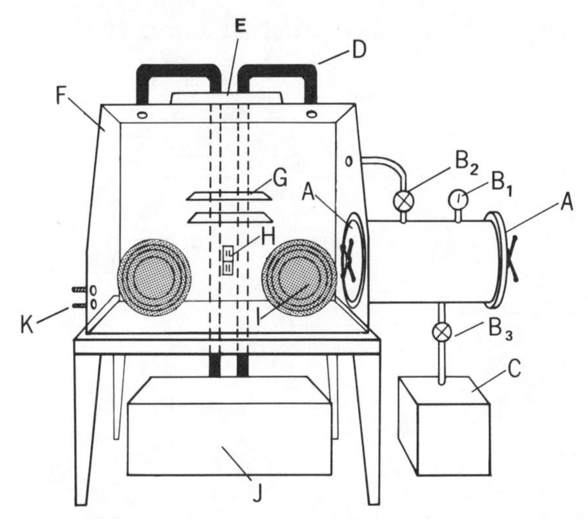

FIG. 5. Inert atmosphere chamber. A, Sealable portals; B_1, pressure gauge; B_2, valve for admitting inert gas to antechamber; B_3, vacuum valve; C, vacuum pump; D, conduits for recirculating chamber atmosphere; E, light; F, box composed of plastic or steel; G, shelves; H, electrical outlets; I, gauntlet; J, gas purification system; K, valves for admitting gases.

[41] See D. F. Shriver, "The Manipulation of Air-Sensitive Compounds," McGraw-Hill, New York, 1969.

might be suitable for the purpose at hand) have at least a slight tendency to leak during a portion of their cycle, it is necessary that the pump be mounted in a chamber that is itself filled with an oxygen- and water-free atmosphere (C in Fig. 6). The pump is housed in a metal cabinet with a bolt-down, gasketed top. The cabinet is *airtight;* it contains five openings other than the top: two for the circulating gas lines that go directly to the pump and out; two for the purge gas, on opposite sides of the box; and a lead-in for the electrical line. The purpose of the pump box is to remove difficulties associated with leakage to or from the pump. Any pump leakage in this setup is harmless, since the box atmosphere is on both sides of the pump. This setup also ensures that the box will not deflate or overfill while the pump is running, although there is sometimes an initial surge when the pump is first turned on. The one-way valve (D) is diagrammed in Fig. 7. It is easily constructed from part K-93125 listed in the catalog of the Kontes Glass Co., Vineland, New Jersey. The pump is an oil-less pressure-vacuum pump which employs carbon rings.[42] Being continuously operable, the pump can still operate satisfactorily after 6 months to 1 year of continuous operation with only slight evidence of wear. It should be noted that an oil-less pump is a necessity. A capacity of from 1.5 cfm to perhaps 3 cfm is satisfactory. Pumps with higher capacity cause problems as a result of pressure buildup

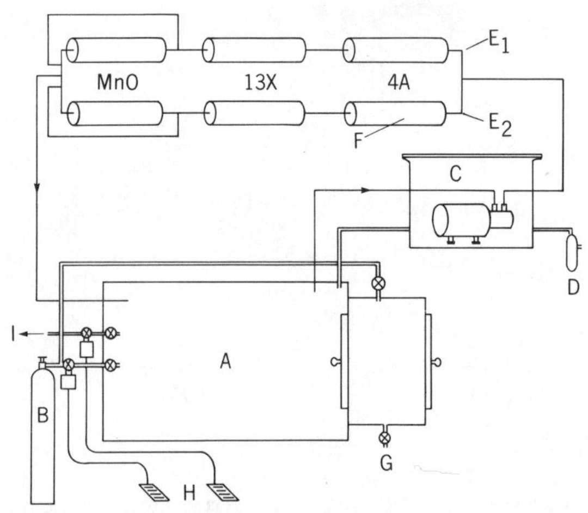

FIG. 6. Inert atmosphere chamber and purification train. A, Inert atmosphere chamber; B, gas cylinder; C, circulating pump; D, one-way valve; E_1 and E_2, alternate purification trains; F, molecular sieves; G, vacuum pump; H, foot pedal-operated solenoid valves; I, excess-pressure valve.

[42] Such pumps are sold by Gelman Instruments Co., 600 S. Wagner Road, Ann Arbor, Michigan 48106; Cole-Parmer Co., 7330 N. Clark Street, Chicago, Illinois 60648; as well as other supply houses.

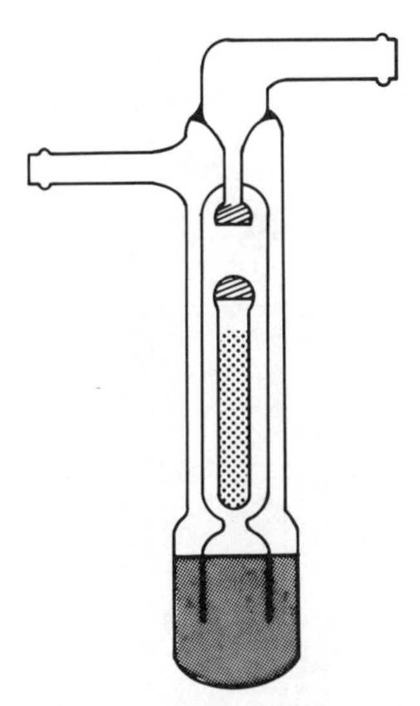

FIG. 7. One-way valve (D) for the pump in Fig. 6.

in the line; pumps with a capacity of less than 1 cfm do not circulate the box atmosphere in a reasonably short period of time.

Column Design. All the columns used in the scavenging system are of Pyrex; the design features are described in Fig. 8.[43] Two parallel-column systems (E in Fig. 6), with provision for switching from one system to the other, have proved convenient. The columns are connected to one another and to the enclosures by $\frac{1}{2}$-in.-i.d. latex rubber tubing with a $\frac{1}{8}$-in. wall thickness. This tubing seems to be relatively impervious to gases and is soft enough so that it can be pinched tightly with standard pinch clamps. Each scavenging line has three glass columns which contain, in the order shown (Fig. 6), Linde 4A molecular sieves, Linde 13X molecular sieves, and manganese(II) oxide on a vermiculite support. The 4A molecular sieves are employed to remove water; 13X molecular sieves are used to remove organic vapors. This is a particularly important feature if oxygen-containing solvents are to be used in the box (e.g., ether); these solvents appear to have a deleterious effect on the manganese(II) oxide columns after a time, and the use of 13X molecular sieves eliminates this problem. The preparation and use of the manganese(II) oxide-containing column is described elsewhere.[43] In practice, the molecular sieve columns are regenerated with about

[43] T. L. Brown, D. W. Dickerhoof, D. A. Bafus, and G. L. Morgan, *Rev. Sci. Instrum.* **33,** 491 (1962). The column design should be changed as follows. Use 1500 g of $Mn(C_2O_4) \cdot 2H_2O$ in each column in 8-cm layers separated by 8-cm layers of vermiculite or other inert support; large-bore stopcocks may be installed at the ends of the columns to replace pinch clamps. Ignore the reference to circulating pump equipment.

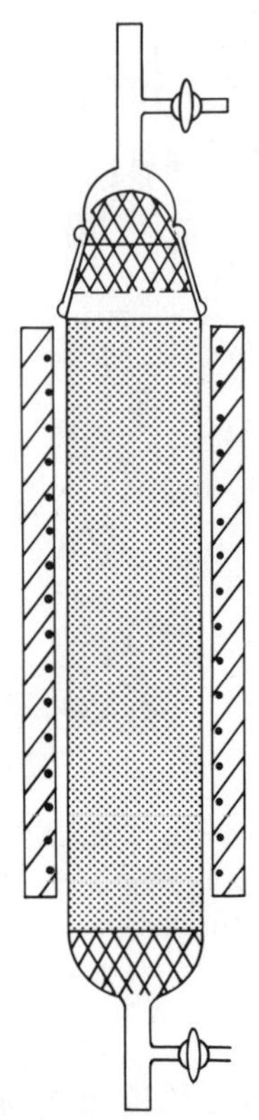

FIG. 8. Deoxygenating column containing manganese(II) oxide. Reprinted with permission from T. L. Brown, D. W. Dickerhoof, D. A. Bafus, and G. L. Morgan, *Rev. Sci. Instrum.* **33,** 491 (1962). Copyright, 1962.)

half the frequency of regeneration of the manganese(II) oxide columns. The bypass line around the manganese(II) oxide column is for use when it is known that there is a substantial amount of ether or other oxygen-containing vapor present in the box atmosphere. The pump is operated for a period of a few hours until the organic vapor is trapped in the molecular sieves.

The key to successful operation of this system is having the pump running *all the time*. The time required for complete removal of water from the box atmosphere seems to be especially long and is not improved materially by using a pump of larger capacity. In doing critical work the best procedure is to place all necessary materials in the main enclosure, making optimal use of the evacuation properties of the transfer port. Then, with all doors tightly closed, a period of at least a few hours is allowed to elapse, with the pump running, before sensitive material is exposed to the box atmosphere.

It is evident that, if the box is to work properly, it must be free of macroscopic leaks. In practice, sufficient pressure is admitted to the box so that the gloves are made to stand out. They should remain in essentially this condition overnight. If they have collapsed against the box within a 12-hr period, the box has an intolerably large leak. The composition of the glove material is of prime importance. The Buta-Sol box glove is the most satisfactory glove that has been used.[44] It exhibits less permeability to oxygen and water vapor than gloves fabricated from other materials. When not in use, the gloves should be kept rolled up and covered with plastic sheeting held in place by an elastic band.

Valves and Other Accessories. It is convenient to have provision for the admission of gas to the box through a solenoid-operated valve controlled by a foot pedal (H in Fig. 6). Any number of solenoid-operated valves (K) that operate with a pressure differential of from 15 to 150 psi are satisfactory. In practice, a 50-lb pressure from the argon tank through a reducing valve is kept on the high-pressure side of the solenoid valve. It is also convenient to have a second solenoid-operated valve leading from the box to a vacuum line or a vacuum pump. This solenoid-operated valve should have a hand-operated valve between it and the box, so that when it is not in use it can be closed off entirely. The second solenoid valve is useful as a means of draining off excess gas from the box while working. The use of these two foot pedals precludes the necessity for bellows or other means of allowing for expansion and contraction of the box atmosphere while the box is being used.

Testing. It is difficult to design a reliable and convenient means of testing the atmosphere for the presence of both oxygen and water. The most convenient and reliable tests are empirical. Thus trialkylaluminum compounds, e.g., triethylaluminum, do not exhibit smoking or other evidence of decomposition on exposure to the box atmosphere when the box is operating properly. Proper operation is evidenced by the length of time required for the manganese(II) oxide columns to turn brown. They should remain green under continuous circulation for at least 2 days if the antechamber has not been used. In the laboratory of R. E. Dessy, Department of Chemistry, Virginia Polytechnic and Commonwealth University, Blacksburg, Virginia, a 25-W light bulb, from which about 1 cm^2 of the glass has been removed by file cuts, is allowed to burn in the chamber atmosphere. It will burn continuously if <1 ppm of water or oxygen is present; it flares out in 5–10 min at higher levels of impurities.

It should be noted that, in putting a box system in operation, a large quantity of purge gas is required for the initial displacement of most of the air. As a minimum, a full-sized 1A tank of argon (B) should be put through

[44] Manufactured by the Norton Co., Safety Products Division, Charleston, South Carolina 29405.

the box; it is also helpful to use a large balloon to expel air. Further, when it is known that contamination of the box has occurred, sufficient time must be allowed for the pump to circulate all of the box atmosphere through the columns to remove the contamination. When a pump of about 1.4 cfm is used, a minimum period of at least 2–3 hr is required for substantial removal of the contamination; it is better to allow about 10–12 hr.

For inert atmosphere chambers without a recirculating purification train, or for those having just recirculation over a drying agent, naturally either a copper metal (Section A,2,a) or the above-described manganese(II) oxide system can be used to purify the nitrogen to flush and fill the antechamber or air lock (Fig. 5). Typically the chemicals and equipment to be introduced are placed in the antechamber, all stoppers and stopcocks being open or wired shut (to avoid explosion upon evacuation). Large, flat-bottomed vessels must be left open. If the design permits, the antechamber is evacuated with an oil pump and refilled with the inert gas from the main chamber. Valve B_2 is left open to the gas supply, and valve B_3 is opened slowly and intermittently. The collapse and extrusion of the rubber gauntlets act as a reservoir during filling of the antechamber. After three or four such cycles, the equipment is admitted into the main chamber. Observance of the following points of technique will ensure continued successful use of an inert atmosphere chamber:

a. If no constant gas purification is provided, a large dish of phosphorus pentoxide and an air blower (if there is an electrical outlet in the box) will minimize moisture.

b. The chamber should be checked for leaks at gauntlets, joints, and gaskets.

c. No wood or paper should be introduced (labels, jar seals).

d. No unnecessary jars or equipment should remain in the box.

e. No volatile organic vapors should be permitted to saturate the atmosphere (to avoid deterioration of plastic walls, gaskets).

f. A shallow metal tray should be present in case of spillage.

g. An electrostatic ground or a radioactive brush should be used to conduct away static charges (to avoid the sticking of powders).

h. Frequently used equipment, e.g., a trip scale, spatulas, weighing glasses, tools, stopper grease, should be kept in the box.

i. Thin-rubber surgical gloves, dusted with talc, should be worn when using the gauntlets; this will protect the gauntlets from undue wear.

3. Inert Solvents[45]

For the preparation of solutions of most organometallic compounds of Groups IA–IIIA the researcher is restricted largely to organic solvents such

[45] See Table I and Appendix III for lists of solvents and their properties.

as saturated and aromatic hydrocarbons, *unreactive* organic halides, ethers, and tertiary amines. Newer or less common systems include zinc alkyls (as a solvent for a Group IA alkyl as a complex[46]), hexamethylphosphoramide,[47] crown ethers,[48] and phosphines.[49] Some of the organometallics of the B families, such as silver and copper(I) acetylides, and the alkyls of zinc, cadmium, and mercury, as well as mixed alkyls [dialkylgold(III) and -thallium derivatives] and organometalloids (*inter alia*, alkyls of germanium, bismuth, and tellerium), are soluble without reaction in a wider array of hydroxylic and polar solvents. Accordingly, solvent purification may be omitted. Nevertheless, despite the inertness of the solvent, one may still have to avoid adventitious air-oxidation (e.g., boron alkyls) or photochemical decomposition (e.g., indium alkyls), both of which may be more serious in solution.

In Section A,1, some of the factors determining the inertness of a given solvent are examined. Here are considered methods feasible for removing small amounts of dissolved oxygen and moisture, as well as reactive structural relatives, from the solvent in question, such as traces of alcohols and peroxides from ether, traces of alkenes or alkynes from alkanes, and *N*-oxides or primary and secondary amines from tertiary amines.

Even with an otherwise pure, dry solvent it is usually advisable to remove any dissolved oxygen gas. Such purification may be achieved in any of the following ways: (*a*) by chilling the solvent (to temperatures that sharply reduce its vapor pressure) and then alternately reducing the pressure above the solvent and restoring the pressure with an inert gas, (*b*) by allowing the inert gas to bubble through the solvent for a time, and/or (*c*) by performing the final distillation of the solvent under nitrogen.

Preliminary chemical extraction of impurities is advisable in the following cases: (*a*) removal of peroxides by ferrous or cerous solutions,[50] (*b*) removal of unsaturates by oxidation with potassium permanganate,[51] (*c*) removal of primary and secondary amines by the use of arylsulfonyl chloride and alkali,[52] and (*d*) removal of phenolic antioxidant additives by alkali—either

---

[46] See F. Hein and F. A. Segitz, *Z. Anorg. Allgem. Chem.* **158,** 153 (1926).

[47] (a) H. Normant, T. Cuvigny, J. Normant, and B. Angelo, *Bull. Soc. Chim. Fr.* 1561 (1965); (b) G. Fraenkel, S. H. Ellis, and D. T. Dix, *J. Am. Chem. Soc.* **87,** 1406 (1965); (c) H. Normant, *Angew. Chem. Int. Ed.* **6,** 1046 (1967).

[48] (a) C. J. Pedersen, *J. Am. Chem. Soc.* **89,** 7017 (1967); **92,** 386, 391 (1970); (b) J. J. Christensen, J. O. Eatough, and R. M. Izatt, *Chem. Rev.* **74,** 351 (1974).

[49] J. Chatt, F. A. Hart, and D. T. Rosevear, *J. Chem. Soc.* 5504 (1961).

[50] J. B. Ramsey and F. T. Aldridge, *J. Am. Chem. Soc.* **77,** 2561 (1955).

[51] E. J. Kuchar, *in* "The Chemistry of Alkenes" (S. Patai, ed.), p. 271, Wiley (Interscience), New York, 1964.

[52] R. Schroter, *in* "Methoden der Organischen Chemie" (E. Müller, ed.), Vol. XI/1, Nitrogen Compounds II, p. 1029. Georg Thieme Verlag, Stuttgart, 1957.

aqueous solutions for water-soluble phenoxides or alcoholic solutions (Claisen's alkali) for those that are water-insoluble.[53]

The commercial availability of solvents of high purity and the ability to examine their homogeneity by gas chromatography or spectroscopy can minimize the need for elaborate chemical extraction treatment in most cases. However, the inevitable presence of peroxides in ethers and unsaturated hydrocarbons demands their preliminary removal. *Attempts to dry peroxide-containing ethers by the addition of hydrides has led to fires and explosions.*[54]

Most organic solvents of reagent grade can be heated directly under reflux with the drying agent. In routine cases, the solvent may be distilled from the drying agent under nitrogen and used directly. In crucial situations the dried solvent should be resubjected to continuous and final drying under nitrogen in the solvent still illustrated in Fig. 9. This apparatus permits freshly dried, deoxygenated solvent to be delivered at will directly into the reaction vessel.

All ground-glass joints should be free of grease and, to prevent their sticking, should be scrupulously clean or be equipped with Teflon or Teflon-coated stoppers. The crucial stopcocks (G) in Fig. 9 have a tendency to stick or leak if they are made of glass. Even if made of Teflon, they can expand upon heating. To deal with the resultant sticking, the Teflon stopcock can be chilled with a piece of solid carbon dioxide before being turned.

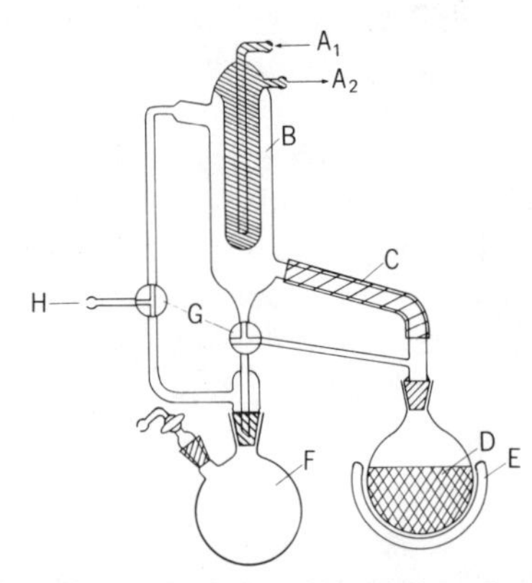

FIG. 9. Still for purifying solvents. A_1 and A_2, Circulation of coolant (preferably kerosene, if Na-K alloy is the drying agent); B, modified Friedrichs condenser; C, asbestos jacketing; D, solvent and purifying agent; E, heating mantle; F, receiver; G, Teflon three-way stopcocks; H, inert gas source.

---

[53] L. F. Fieser and M. Fieser, "Reagents for Organic Syntheses," Vol. 1, p. 153, Wiley, New York, 1967.

[54] R. B. Moffitt and B. D. Aspergren, *Chem. Eng. News* **32,** 4328 (1954); J. B. Wommack, *ibid.* **55,** 6 (1977). For a survey of peroxide removal methods, see J. A. Riddick and W. B. Bunger, "Organic Solvents," 3rd ed., p. 690. Wiley (Interscience), New York, 1970.

The recommended purification procedures for nonhydroxylic solvents are described below according to increasing periodic group number of the key element.

Group II. The use of zinc alkyls as Lewis-acid coordinating solvents for alkali metal alkyls has proved valuable in physicochemical studies, such as research on electrolysis, complex ion stabilities, and spectroscopic measurements of carbanions. Zinc alkyls can now be synthesized from commercially available aluminum alkyls and anhydrous zinc chloride.[55] The resulting dialkylaluminum chloride can be complexed with potassium chloride, and the zinc alkyl distilled away. The relatively pure zinc alkyl is then heated with powdered anhydrous potassium fluoride, which forms complexes with traces of R_3Al and R_2AlCl (i.e., $K[R_3AlF]$ and $[R_2AlF]_2$), and then the zinc alkyl is fractionally distilled under nitrogen.

Group IV. The principal hydrocarbons of interest as solvents are alkanes, cycloalkanes ($C_n > 4$), and aromatics. Volatile saturated hydrocarbons often may be freed of traces of unsaturated hydrocarbons by heating with small amounts (1 ml/100 ml) of commercially available diethylaluminum hydride. At the same time residual oxygen and moisture can be scavenged. A final distillation from this solution provides a pure saturated hydrocarbon.

Alternatively, for saturated hydrocarbons or aromatics, the solvent may be refluxed for 24 hr over molten sodium, liquid sodium amalgam, or powdered calcium hydride. After direct distillation the distillate can be redistilled immediately, after treatment with lithium aluminum hydride, from the still pictured in Fig. 9.

Before use, it may be advantageous to test any stored solvent for peroxides (see below), for unsaturates,[51] and for replaceable or "active" hydrogen (Section D,4).

Group V. Commonly, only tertiary amines, phosphines, and arsines are acceptable solvents in this family, although *N,N*-dimethylformamides and acetonitrile may be of use with less reactive metal alkyls. However, even the tertiary amine, pyridine, has been found to undergo carbon–metal bond addition when employed as a presumably inert solvent with lithium alkyls.[56]

Tertiary amines can be dried by distilling from powdered phosphorus pentoxides and then from a suspension with lithium aluminum hydride. The latter treatment should serve to remove traces of partially alkylated amines. Such operations should be conducted under dry nitrogen directly before use, as amines tend to absorb water and carbon dioxide from the air.

The hydrophobic character of tertiary phosphines and arsines means that

[55] K. Ziegler and E. Hüther, U.S. Patent 3,124,604 [*Chem. Abstr.* **60,** 15909 (1964)].

[56] K. Ziegler and H. Wollschitt, *Justus Liebigs Ann. Chem.* **479,** 123 (1930); K. Ziegler and H. Zeiser, *ibid.* **485,** 174 (1931).

a treatment similar to that for amines will be more than adequate to remove water. Yet the ease with which these compounds oxidize in air to R_3EO means that a nitrogen atmosphere must be maintained at all times.

A test of the resulting R_3E products for active hydrogen is recommended.

Group VI. Ethers of various structural types are the most common solvents for the preparation and subsequent reactions of organomagnesium and organolithium reagents. Less commonly, sulfides may be occasionally employed. Anhydrous and deoxygenated ethers may be secured by simply adding a portion of a preformed organometallic (phenylmagnesium bromide or triethylaluminum) and distilling off the volatile ether (e.g., diethyl ether). More routinely, the ether is heated at reflux and distilled from freshly pressed sodium wire or sliced pieces. The fresh distillate may be subjected to a final distillation after the addition of lithium aluminum hydride (*caution:* check for, and remove if necessary, peroxides before adding hydrides[54]) or benzophenone and sodium metal [which lead to the formation of soluble ketyl, $(C_6H_5)_2\dot{C}—O^- Na^+$]. Certain ethers will undergo cleavage upon exposure to alkali metals or their organic adducts[57] (diallyl ether, diphenyl ether, etc.), and the alternative use of added R_3Al may be advisable.

Before and after purification the ethers should be tested for peroxidic impurities,[58a] either by a starch–iodide[58b] or by a vanadium(V) procedure.[59] The former method succeeds best if a starch solution is added to a mixture of 1 ml each of solvent, saturated aqueous potassium iodide, and glacial acetic acid (blue color for a positive test). In the latter method a 0.2% solution of vanadium pentoxide dissolved in 3.5% sulfuric acid indicates the presence of peroxides at concentrations as low as $3.5\% \times 10^{-4}$ *M*. Since the best tests are those capable of being placed on a quantitative basis, the following modified iodide procedure[58b] is recommended for organic solvents.

A 5-ml sample of solvent is admixed with 20 ml of isopropyl alcohol and 1 ml of glacial acetic acid. After heating to reflux, 5 ml of a saturated solution of sodium iodide in isopropyl alcohol is added and the mixture heated at reflux for 5 min. A 5-ml volume of water is introduced, and the solution is titrated with a 0.01 *N* aqueous solution of sodium thiosulfate until the disappearance of the yellow color.[60,61]

---

[57] (a) P. Schorigin, *Ber.* **56,** 176 (1923); (b) K. Ziegler and F. Thielmann, *ibid.* **56,** 1740, 2453 (1923); (c) H. Gilman and J. J. Dietrich, *J. Org. Chem.* **22,** 851 (1957); (d) H. Gilman and J. J. Dietrich, *J. Am. Chem. Soc.* **80,** 380 (1958); (e) J. J. Eisch, *J. Org. Chem.* **28,** 707 (1963).

[58] (a) J. A. Dukes, *United Kingdom Atomic Energy Authority, Ind. Group Hdq.* D1 (S) TN-2010 (1951) [*Chem. Abstr.* **53,** 11194 (1959)]; (b) C. D. Wagner, R. H. Smith, and E. D. Peters, *Anal. Chem.* **19,** 976 (1947).

[59] A. Valseth, *Medd. Farm. Selskap* **15,** 21 (1953) [*Chem. Abstr.* **47,** 11655 (1953)].

[60] A spectrophotometric modification of this method detects active oxygen in the range of 5–80 ppm [D. K. Banerjee and C. C. Budke, *Anal. Chem.* **36,** 792 (1964)].

[61] P. R. Dugan [*Anal. Chem.* **33,** 1630 (1961)] suggests the use of *N,N*-dimethyl-*p*-phenylenediamine sulfate for the detection of peroxides.

Peroxides can be removed from organic solvents by treatment with cerium(III)[50] or iron(II)[59] salts. Thus cerium(III) hydroxide may be prepared by treating an aqueous solution of cerium(III) chloride with sodium hydroxide until the supernatant solution is slightly alkaline. The precipitated white hydroxide can be isolated and washed by centrifugation but should not be dried. The solvent to be purified should be shaken with the moist cerium(III) hydroxide for 15 min, during which time the solid will turn reddish-brown if peroxide is present. With water-immiscible ethers about 2 ml of water will speed the removal of peroxides by the cerium(III) hydroxide.

Alternatively, 1-liter portions of the ether should be shaken for 15 min with 200 ml of a 10% solution of sulfuric acid containing 20 g of iron(II) sulfate (as the heptahydrate). The ether layer then is decanted onto 30 g of potassium hydroxide pellets and stored for 1 hr with periodic shaking.[62]

The ether, as well as other purified solvents, may be tested for active hydrogen (OH, NH, certain CH) in the following way: Adapter D of the apparatus shown in Fig. 10 is replaced by a septum closure, and a sample

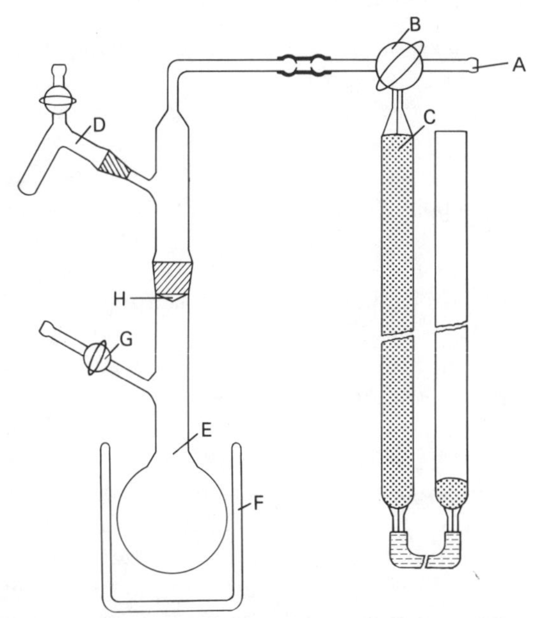

FIG. 10. Gasometric apparatus for the protolysis of organometallics. A, Inert gas; B, three-way T stopcock; C, mercury-filled buret; D, turnable tube containing the proton source; E, vessel containing the organometallic which is introduced by syringe or in a tared, crushable vial; F, bath; G, three-way stopcock; H, joint with a drip tip.

[62] W. Dasler and C. D. Bauer [*Ind. Eng. Chem. Anal. Ed.* **18,** 52 (1946)] recommend the passge of a peroxide-containing solvent through a column of activated alumina. Of great advantage is that the eluate is at once free of peroxides and is essentially anhydrous. See G. Wohlleben [*Angew. Chem.* **67,** 741 (1955)] for the drying of solvents with activated alumina.

of the purified solvent is injected into the apparatus. Then via an inserted syringe a solution of lithium aluminum hydride in dry THF is added dropwise. Any significant effervescence indicates that the solvent contains active hydrogen groups.[63]

Group VII. Although monohalides, such as chlorobenzene, are sometimes used as inert solvents, polyhalides such as carbon tetrachloride, chloroform, and methylene chloride are only suitable for less reactive organometallics and organometalloids. Such halides should be washed well with water to remove acidic and alcoholic additives and dried over anhydrous calcium sulfate. Final drying and distillation from molecular sieves or powdered phosphorus pentoxide are recommended. *However, in no case should alkali metals, reactive metal alkyls (RLi, R_2Mg, or R_3Al), metal hydrides, or strong alkali be used to dry halogenated hydrocarbons. Severe decomposition or an explosion may result.*[64,108]

4. Drying and Deoxygenating the Reaction Zone

Scrupulous purification of the inert gas and of the solvents should be accompanied by a similar preparation of all parts of the reaction vessel and of the reactants themselves. Depending upon the variety of operations to be performed in the vessel (e.g., reaction and distillation), the apparatus should be provided with a minimum of ground-glass joints. Such connections constitute a source of possible decomposition through air leakage, frustration through freezing, and contamination through grease extraction. If joints must be used, they should be treated sparingly over the upper half of the joint with silicone grease and the parts worked together to distribute the coating. Glass hooks on either side of the joint should be provided with spiral springs. A pouring spout on the inner joint member serves to minimize grease extraction (Fig. 11). Finally, in certain cases, standard joints and stopcocks of Teflon may be useful where grease contamination must be excluded completely.[65] Where contact with solvent vapor is not a serious concern, joints or other glass tubing apertures may be provided with rubber or plastic septa, whose resiliency permits samples to be added or removed with needle syringes without loss of inert atmosphere protection.

If a shaft stirrer is employed, the bearing seal becomes a problem of concern. Various all-glass stirring assemblies are available commercially, whose

---

[63] See N. G. Gaylord, "Reduction with Complex Metal Hydrides," pp. 76–82, Wiley (Interscience), New York, 1956, for another useful form of a gasometric apparatus.

[64] H. Staudinger, *Z. Angew. Chem.* **35,** 657 (1922).

[65] Teflon stopcock plugs fitting smooth-bore or ground joints are available from several suppliers. Such plugs can be prevented from sticking by rotating them in the bore during heating or cooling operations (see above for the freeing of stuck joints).

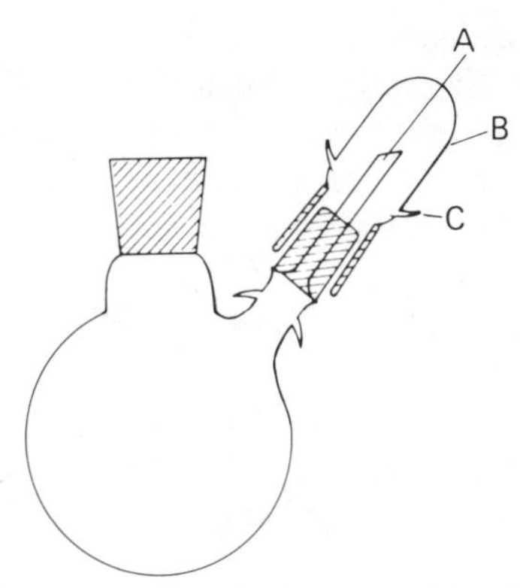

FIG. 11. Fused-on pouring spout and cap (spout can be curved if the cap is of suitable shape): A, pouring spout; B, cap; and C, steel spring anchors.

precisely ground bearing and shaft can maintain pressures down into the micrometer range. However, misuse and wear eventually require that a viscous, inert lubricant be used to maintain an inert atmosphere in the vessel. Bearing cooling (using water or another liquid) is mandatory. Alternatively, a sealed stirrer of the magnetic coupling or bar variety or one equipped with a Teflon-clad ball-joint seal[66] can be employed if no contamination is permitted (see other types in Fig. 12).

Once the foregoing union of joints has been considered, all glassware is heated overnight in an oven at 200°. The apparatus is assembled as rapidly as possible while still warm. Via the inert gas distribution manifold, the

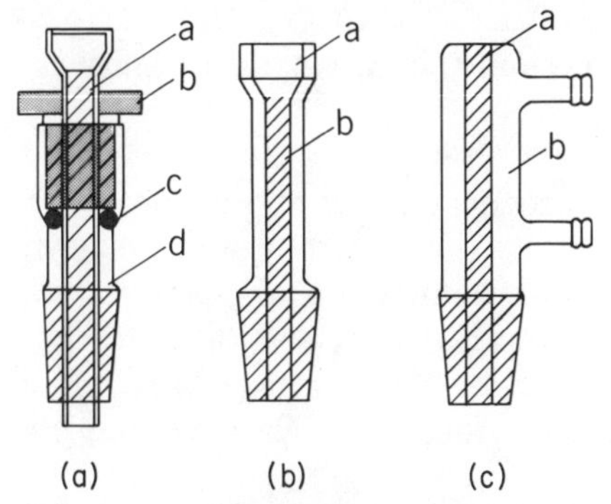

FIG. 12. Types of seals for a motor-driven blade stirrer. (a) Adjustable nylon bushing with an O-ring seal (a, ground aperture for stirring shaft; b, adjustable screw; c, ring seal; d, outer joint) (Ace Trubore); (b) precision-ground bearing to fit the shaft, with a lubricant cup (a, lubricant cup; b, ground aperture) C, water-jacketed, precision-ground bearing (a, ground aperture; b, water jacket).

[66] Ace Glass, Inc., 1430 Northwest Boulevard, Vineland, New Jersey 08360.

the whole apparatus is evacuated (*caution:* avoid thin-walled, flat-bottomed parts) and "flamed," as described in Section A,2,c. If heat-insensitive solid reactants are to be introduced into the vessel, they can be included in the flaming. In cases where the apparatus has been charged with reactants in the dry box, the flaming and evacuation are omitted and the system simply purged freely with the inert gas.

However, all solid, liquid, and gaseous reactants, even if oxygen-insensitive, should be dried thoroughly. Stable inorganic salts should be ground thoroughly (see Section B) and heated at 300–400° and allowed to cool under nitrogen. Otherwise pure organic solids should be stored over phosphorus pentoxide in a desiccator or heated in an Abderhalden drying pistol charged with this dessiccant, employing a solvent whose boiling point will cause the solid neither to melt nor to sublime.

The drying of liquid reactants[67] naturally requires a greater variety of methods than those suitable for inert solvents. Many alkyl halides can be dried by storage over phosphorus pentoxide, while olefins and internal acetylenes can be distilled from lithium aluminum hydride. In all cases, fractional distillation under nitrogen in a dry apparatus is recommended. Gas chromatographic and spectral checks for isomerization or decomposition should always be made.

The most common gaseous reactants in organometallic procedures are carbon monoxide, carbon dioxide, oxygen, hydrogen, and lower hydrocarbons. The passage of inorganic gases through a drying tower of phosphorus pentoxide should be satisfactory (Section A,2,b); specialized purification procedures should be consulted for organic gases.[68]

5. Storage and Transfer of Organometallic Compounds

a. Storage

Storage of pure research samples of organometallics for future use must be in sealed glass tubes. For this purpose a Schlenk tube[69] of suitable capacity, equipped with 6-mm-i.d. glass apertures at a 60° angle, is thoroughly dried and filled with an inert gas (Fig. 13). With the aid of a modified Ziegler trough,[70] a siphoning arrangement (Fig. 14), a solids transfer tube (Fig. 15), or a gastight syringe (Fig. 32a), the sample is introduced into leg

[67] J. A. Riddick and W. B. Bunger, "Organic Solvents," 3rd ed., Wiley (Interscience), New York, 1970.

[68] H. Melville and B. G. Gowerlock, "Experimental Methods in Gas Reactions," Macmillan, New York, 1964.

[69] W. Schlenk and A. Thal, *Chem. Ber.* **46,** 2843 (1913).

[70] K. Ziegler, H. G. Gellert, H. Martin, K. Nagel, and J. Schneider, *Justus Liebigs Ann. Chem.* **589,** 119 (1954).

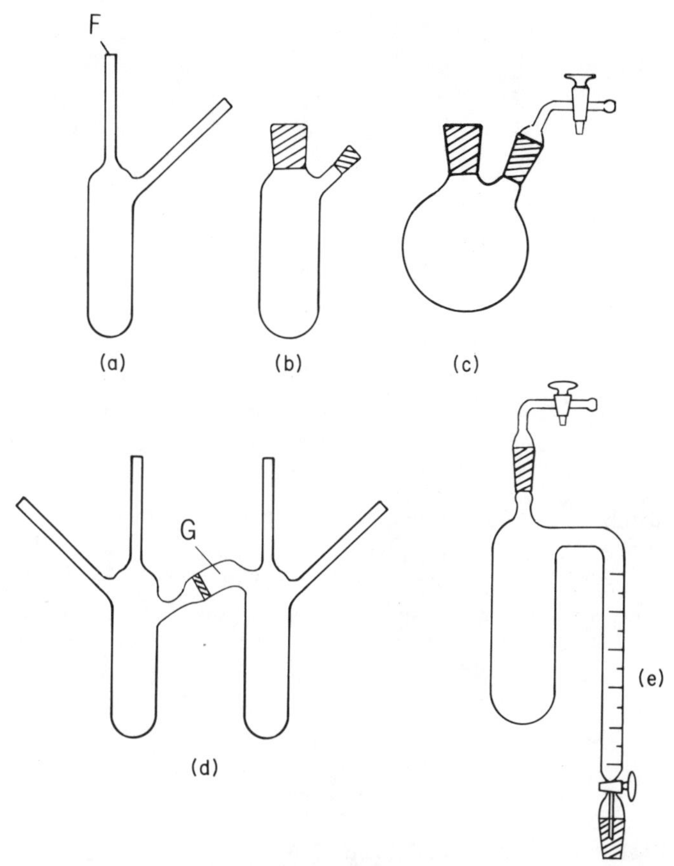

FIG. 13. Variants of Schlenk vessels. (a) Classic type for fusion closure (F); (b) centrifuge vessel; (c) reusable storage or reaction vessel with a three-way stopcock adapter; (d) double tube for filtration through frit G; (e) tube equipped with a fused-on buret for dispensing and storing reagent solutions.

F of the Schlenk tube (Fig. 16). Care should be taken not to allow the liquid sample to touch the walls of the tube. With the nitrogen vacuum adapter on leg F (Fig. 16) a stream of nitrogen is allowed to pass out of leg E while the contents of the tube are cooled with ice, solid carbon dioxide, or another bath.

With a glassblower's torch the middle of tube E is warmed gradually, uniformly and thoroughly, while nitrogen is maintained. At a rate for maintaining glass wall thickness the upper end of the tube is grasped with forceps and slowly and steadily drawn out to a fine-bore capillary several millimeters in length. The inert gas stopcock at D is then turned to stop the main gas flow to the vessel, and the Schlenk tube capillary is then sealed off quickly

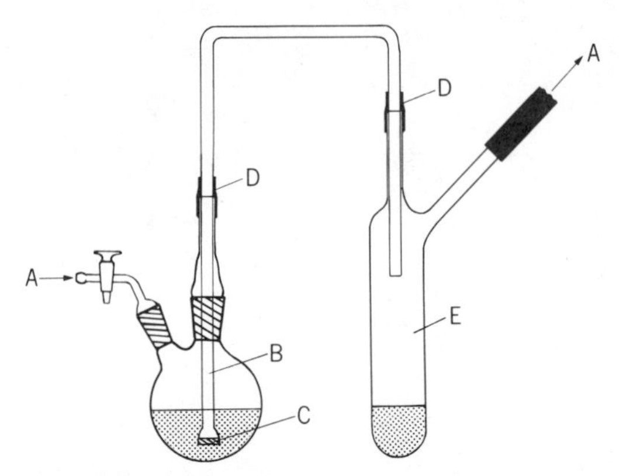

FIG. 14. Siphon transfer or filtration of liquid organometallics. A, Inert gas source; B, glass siphon; C, fused-on frit (omitted when simple transfer without filtration is required); D, rubber tubing to provide a snug but mobile contact; E, receiving vessel.

with the torch. The tip of the seal-off point can be annealed to thicken and strengthen the glass. A similar procedure is employed to seal tube F. It is important not to heat too wide an area, but only the tube tip, especially with F, since no pressure release is possible. Upon cooling, the sealed tube

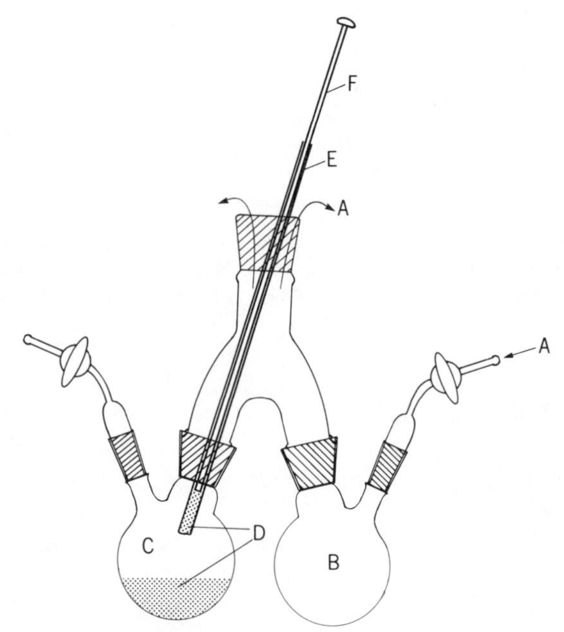

FIG. 15. Device for transferring sensitive organometallic solids. A, Inert gas current; B, receiver; C, storage vessel; D, solid to be transferred; E, glass tube; F, glass rodding fitting E snugly.

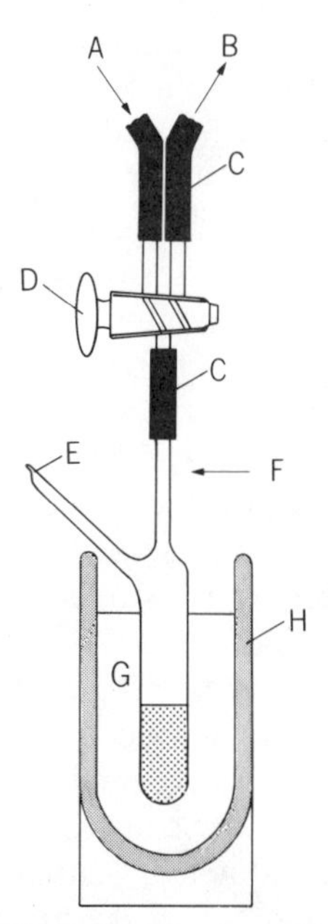

FIG. 16. Setup for sealing Schlenk tubes. By alternately turning D to A (inert gas) or B (vacuum), the pressure in F can be kept normal or reduced. Tube E was melted closed by a torch while being cooled with bath G in Dewar H. Tube F is now melted shut while under normal pressure of inert gas (C, heavy-walled rubber tubing).

should be inspected carefully for the adequacy of the seals. The sample should be labeled securely as to date, origin, nature, and physical constants. Furthermore, it is often worthwhile to paint the vessel with an opaque material to exclude light (leaving a coverable window for inspecting the contents) or at least to envelope it in metal foil. Such vessels should be kept upright in a dark place at a moderate temperature. To protect from shock and the consequences of breakage, these vessels should be placed in metal troughs of ample capacity filled with an inert absorbent (diatomaceous earth or calcium carbonate). These troughs should be *stored away from combustibles.*

In the case of solids, a Schlenk tube like that in Fig. 13 is equipped with rubber-connected stopcocks and then charged with the solid organometallic in the dry box. Alternatively, an inert atmosphere chamber devised from a multientry desiccator (three or more holes in the cover; see Fig. 33) or a collapsible polyethylene bag[35] may be employed. As with liquids, the solid should not contact the walls of the entry tube. To this end, entry tube F may be made of wider bore (15 mm) to accommodate a long-stem solids funnel; or a glass tube of 4–5 mm o.d. may be equipped with a snugly fitting glass cane (Fig. 15) and the unit employed as an augerlike devise. The glass tube is filled by packing with solid, the outside freed of particles, the tube placed through F to the bottom of the Schlenk tube, and the cane plunger used to extrude the solid. Flushing and sealing the tube can then be performed as with liquid samples.

The storage of periodically used organometallic reagents does not demand such elaborate techniques as those described above. However, certain operating procedures given below may contribute to trouble-free work. Organometallics of a solid or liquid character may be kept for a period either in an inert atmosphere chamber (dry box) or in a vessel designed for the simultaneous introduction of inert gas and sample withdrawal. The types of vessels suitable for this purpose are a modified Schlenk vessel (Fig. 13c) and a bottle or flask equipped with a serum spetum (Fig. 17). In the former, the inert gas may be introduced at the stopcock and the stopper then slowly raised under positive pressure. The three-way stopcock[71] permits the removal of any air in the gas tube before the inert gas is admitted to the vessel. With septa apertures the needle of an inert gas hose lead first is inserted through the septum, and then a needle syringe, preflushed with inert gas, is inserted for sample removal (Fig. 33). With Schlenk tubes bearing ground joints the silicone grease frequently is attacked, leading to leaks or frozen joints. Naturally, frozen stoppers and stopcocks are a hazard and a frustration which cannot be alleviated with standard ways for freeing stuck joints. Since prevention is more feasible, all joints should be kept from contacting the

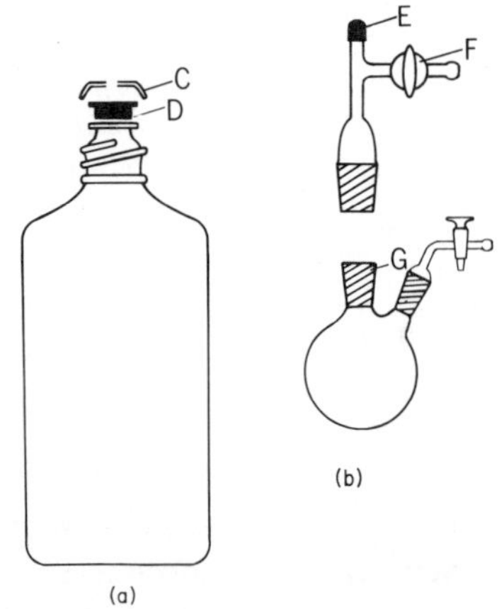

FIG. 17. Storage vessels for organometallics. (a) Commercial bottle fitted with a pierceable septum (D) and a pressure cap (C) having a hole; (b) Schlenk vessel with a three-way stopcock and bearing a stopper at G (storage) or a septum-bearing (E) adapter (F).

[71] Design employed in the Max-Planck-Institut für Kohlenforschung, Mülheim (Ruhr), Federal Republic of Germany.

organometallic directly and, under nitrogen, should be wiped clean and regreased frequently. If the researcher is ever presented with frozen joints in a supply of metal alkyls of Groups IA–IIIA, we advise him to view the vessel as a napalm grenade and have it disposed of in an open-air dump. Hardy researchers may wish to clamp the vessel upright in a metal trough and to apply warm copper bands to the frozen joint. A safety shield, impermeable asbestos or leather gauntlets, and a face shield are mandatory for handling spontaneously flammable samples.

Vessels with septa entry usually are encountered with reagents from chemical suppliers (Appendix II), with stock solutions prepared in an inert atmosphere chamber, or with reaction vessels for kinetic and polymerization studies. When not in use, reagents with punctured septa are best resealed or stored in the dry box. If possible, the septa should be replaced frequently.[72]

b. TRANSFER

Larger quantities of liquid metal alkyls (boron, aluminum) are furnished in steel cylinders.[73] Although the special instructions accompanying such cylinders should be adhered to closely, the following constitutes a safe, useful procedure for the single-exit storage cylinder. A Hoke needle valve[74] (Fig. 18) fitting the threading on ordinary lecture bottles and larger cylinders is soldered or brazed to a 25-cm length of 10-mm-o.d. copper tubing. The storage cylinder is inverted with secure clamping, and the needle valve and tubing setup are attached as shown in Fig. 18. (Be sure the Hoke valve is tight.) The exit copper tube passes through a reducing adapter into a prepared Schlenk storage vessel of suitable capacity. A section of rubber tubing between the copper tube and the adapter permits flexibility and completes the seal. *A metal trough filled with diatomaceous earth must be provided in case of leakage or breakage.* Via C, the whole apparatus is evacuated and refilled with inert gas (with the Hoke needle valve open and the cylinder valve shut). At normal pressure under the inert gas, the Hoke valve is closed and the cylinder valve opened. Opening of the Hoke valve now permits gravity flow of the organometallic reagent into the vessel. After the proper amount of delivery, the main cylinder valve is closed and the Hoke valve

---

[72] Special septa for various apertures may be purchased from the Aldrich Chemical Co., Milwaukee, Wisconsin 53233.

For an excellent discussion of liquid transfer techniques involving hypodermic syringes, see G. W. Kramer, A. B. Levy, and M. M. Midland, *in* "Organic Syntheses via Boranes" (H. C. Brown, ed.), Chapter 9, Wiley, New York, 1975.

[73] See the bulletin, "Aluminum Alkyls," distributed by Texas Alkyls, Inc. and Stauffer Chemical Company, 1976, for a description of proper handling, analytical techniques and safety precautions.

[74] Hoke needle valves and stainless steel reaction vessels (bombs) are available from Hoke, Inc., Tenakill Park, Cresskill, New Jersey.

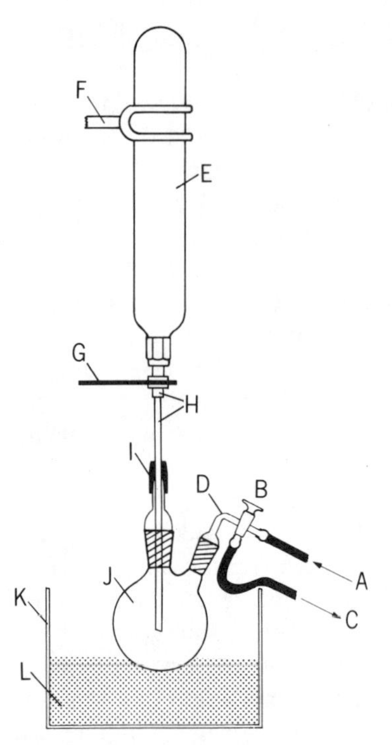

FIG. 18. Arrangement for transferring organometallics from small, single-entry cylinders. A, Inert gas; B, three-way stopcock; C, vacuum; D, stopcock adapter; E, storage cylinder; F, clamp support; G, valve wrench (in this case, Hoke valve on H is omitted); H, copper tube with an adapter fitting the cylinder opening; I, rubber tubing for a snug contact; J, receiver; K, large metal trough; and L, inert absorbent (vermiculite).

opened completely, permitting drainage of the copper tube. Cautiously the storage vessel is lowered to extricate the copper tube. (*Danger: because of droplets, wear asbestos gloves.*) The adapter is removed, and a freshly greased stopper inserted. Such a stopper should not be jammed tight until the stopcock at B is turned off. Overpressure is thereby avoided. Stopcocks and stoppers should be wired shut. After standing some time, the Hoke valve setup is cautiously removed from the cylinders (using gloves). After rinsing with benzene, the tube can be washed in turn with benzene–methanol, water, and dilute hydrochloric acid. The valve setup should be dried carefully with warming (*caution: solder*) before reuse.

The transfer of smaller amounts of reagents, of solutions of reaction products, or of pure products involves, in part, techniques concerned with those of Section C, but here certain general aspects will be treated. Reaction solutions can be withdrawn from the site of reaction by gravity filtration or by decantation. Filtration is achieved most conveniently for Grignard reagent or organolithium preparations by use of the reaction vessel[75] illustrated in Fig. 19. At the end of the preparation the solution may be separated from insoluble metal or salts by withdrawing it through the coarse frit directly into a storage or reaction vessel. On the other hand, decantation through a plug of glass wool using the vessel shown in Fig. 11 may be equally effective for such solutions.

Techniques of transfer suitable for even the most air- and moisture-sensitive samples include siphoning, pipetting, and volatilization. A siphon having unequal legs (Fig. 14), namely, a piece of bent glass tubing of 2 mm i.d., can be mounted via an adapter and seal of rubber tubing (D) to vessel B. At first (position 1) the siphon leg in B is held above the liquid, and it is swept out with the inert gas. Then the shorter leg of the siphon is introduced into a preflushed storage or reaction vessel (E). Finally, both legs of the siphon are

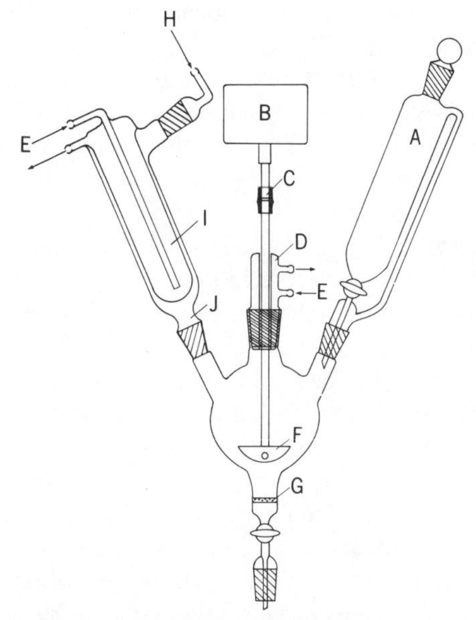

FIG. 19. Reaction apparatus for preparing and filtering organometallic solutions. A, Pressure-equalized addition funnel (graduated); B, stirring motor (sparkless); C, flexible connection; D, precision-ground bearing with cooling (E); F, blade stirrer; G, glass frit; H, inert gas ingress; I, Friedrichs condenser; J, shortened condenser shank to keep the ring of condensation below the stirrer bearing.

[75] The reaction vessel may be fashioned with a glass cooling mantle or, alternatively, the bottom of a styrofoam container can be bored out to fit the vessel up over the stopcock. A water-insoluble cement can be used to join the container and the reaction vessel. A spigot in the side of the styrofoam mantle can be used to draw off water (C. A. Kovacs, The Catholic Univ. of America, Washington, D.C., 1971).

lowered until the longer end in B is beneath the liquid (position 2). By pressure of the inert gas in vessel B the liquid is forced over into E. At the end of delivery the siphon is raised out of liquid in B, and usually the liquid in the siphon drains out. By fusing a filter frit (C) to the siphon, this procedure can be adapted to filtering off and washing reaction solids.

Smaller amounts of measured liquids can be transferred with pipets fused to a syringe (Fortuna pipets[76]) or connected to 50- or 100-ml syringe plungers by lengths of rubber tubing, or with airtight syringes.[77] Such pipets or syringes are flushed by alternately inserting the tip into an inert gas lead and drawing the plunger out partially, and then removing the pipet tip and expressing the gas. At the last filling, the plunger is left extended, and the gas expressed just before use. Although small syringes can be flushed similarly and be used for microsamples, the plunger must be gastight to avoid contamination.[77] It should be noted that the use of plunger pipets permits the plunger to be greased (with ordinary stopcock petroleum base) without contamination of the organometallic sample.

The withdrawal of samples of pure organometallic liquids or solids often can be accomplished by a modified volatilization process, such as distillation or sublimation (Section C,6). In a related fashion, samples for various analytical or storage purposes can be collected by use of a pressure differential.[78] For liquids, ampuls of either thin-walled or thicker-walled glass (Fig. 20) are inverted on the wire loop of the adapter[79] and by adjustment of the movable glass tube are held above the liquid. Through stopcock I the pressure in the vessel is reduced. Then the adjustable tube is lowered until just the tip of the ampul is immersed in the liquid. Then inert gas is admitted slowly to force the desired amount of liquid up into the ampul. The ampul is taken out of the liquid, and ordinary pressure restored. If the capillary of the ampul is thin and rather small, exposure to air will lead to sufficient decomposition to plug the aperture and protect the sample, while the upright ampul is sealed with a microflame. In other cases of immediate further use, the amount of decomposition at the tip may be so trivial as to provide a weighed sample for analysis by taring the ampul empty and full. This same goal can be achieved with solid samples by employing ampuls with tightly fitting ground stoppers[80] (Fig. 15). By means of the wire loop these ampuls can be suspended from the beam of the analytical balance or held in the extended neck of the vessel whose contents are to be sampled.

---

[76] W. Strohmeier, *Chem. Ber.* **88,** 1220 (1955).

[77] A variety of airtight syringes is available from the Hamilton Co., Whittier, California 90608.

[78] E. Krause and A. von Grosse, "Die Chemie der Metallorganischen Chemie," p. 806, Verlag Gebr. Borntraeger, Berlin, 1937.

[79] See G. Bähr and G. Müller, *Chem. Ber.* **88,** 251 (1955).

[80] H. Schlesinger and H. Brown, *J. Am. Chem. Soc.* **62,** 3433 (1940).

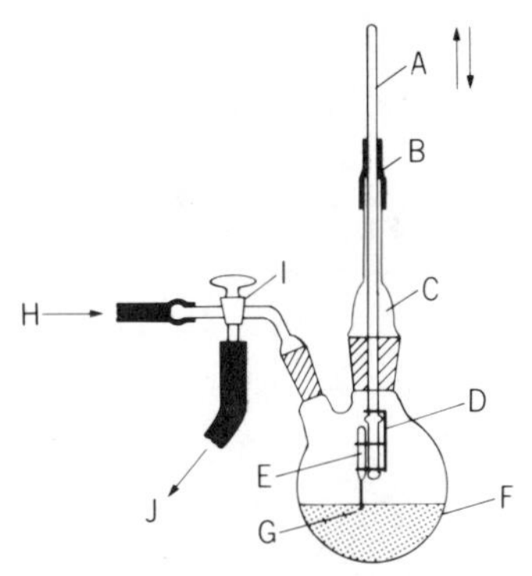

FIG. 20. Technique for filling ampuls with liquid organometallics. A, Movable hollow glass tube bearing a wire support (D) for inverted glass ampuls (E); B, rubber tubing for a snug, flexible contact; C, adapter whose upper tube fits A closely; F, liquid to be sampled; G, when immersed, ampul should be just below the surface; H, inert gas source; I, three-way stopcock; J, vacuum.

The use of a solids auger or a long-shafted spatula scoop can effect the ampul filling. The glass cap is put on, and the ampul is withdrawn and retared.

Where the setup involved permits the introduction of all vessels into an inert atmosphere chamber, the foregoing sampling procedures can be simplified. Since the evacuation of closed glass vessels in the air lock of such a chamber is time-consuming and hazardous (glassware may be shattered by ejected stoppers), the above time-tested techniques are not to be scorned. Indeed, except for the most elaborate inert gas chambers (those having constant recirculation and purification), the atmosphere control in these techniques is probably as good as or better than those involving a dry box.

B. Execution of Organometallic Reactions

1. Purity of Reagents

The most common starting materials for organometallic preparations are metals, preformed or commercially available organometallic compounds (see Appendix II for a list of commercial suppliers of organometallic compounds), metal salts, elemental gases (H_2, CO, etc.; Section A), organic halides, ethers, and unsaturated hydrocarbons. Unless experience has verified that a given grade of starting material gives consistently satisfactory results, both inorganic and organic reagents should be certified as to purity and as to the kinds of trace impurities present. This caveat is essential, if high yields of pure organometallics are to be obtained. Moreover, even if the impurities do not interfere in the preparation of the organometallic itself, their presence may provoke abnormal results in subsequent reactions. Such impurities are traces of transition metals in Grignard solutions, which at once confound and excite researchers who encounter the resulting abnormal reactions. The classic observations by Kharasch and co-workers[81] [e.g., Eq. (6)][82] have been augmented by many recent discoveries, such as the finding by Finkbeiner and Cooper[83] that traces of titanium salts in magnesium can promote a hydride transfer reaction between Grignard reagents and olefins [Eq. (7)]:

$$(C_6H_5)_2\underset{\text{OH}}{C}-\underset{\text{OH}}{C}(C_6H_5)_2 \xleftarrow[2\%\,CoCl_2]{CH_3MgBr} (C_6H_5)_2C{=}O \xrightarrow{CH_3MgBr} (C_6H_5)_2C\genfrac{}{}{0pt}{}{OH}{CH_3} \qquad (6)$$

$$C_6H_5CH_2CH{=}CH_2 \xrightarrow[1\%\,TiCl_4]{(CH_3)_2CHMgBr} C_6H_5CH_2CH_2CH_2MgBr + CH_3CH{=}CH_2 \qquad (7)$$

Besides the profound effect transition metals can have on the behavior of magnesium and the resulting Grignard solutions, the reaction of aluminum

---

[81] M. S. Kharasch and O. Reinmuth, "Grignard Reactions of Nonmetallic Substances," pp. 117–131, Prentice-Hall, Englewood Cliffs, New Jersey, 1954.

[82] M. S. Kharasch and F. L. Lambert, *J. Am. Chem. Soc.* **63,** 2315 (1941).

[83] (a) H. L. Finkbeiner and G. D. Cooper, *J. Org. Chem.* **26,** 4779 (1961); (b) H. L. Finkbeiner and G. D. Cooper, *ibid.* **27,** 3395 (1962).

metal with hydrogen and olefin to form aluminum alkyls (Ziegler processes[84]) is known to be accelerated by certain transition metals.[85] Likewise, a small amount of sodium in commercially available lithium is advantageous for the ease of organolithium preparations.[86]

Aside from the inherent purity of a metal, the surface of the suitably subdivided metal (Section B,2) should be free of organic matter and any coating. For soft reactive metals (Li, Na) the metal may be pressed as wire directly into the reaction vessel. Alternatively, chunks of metal may be immersed in dry benzene or xylene in a Petri dish and scraped free of oxide or nitride (Li). Extraordinary caution should be taken in handling potassium metal and its scrapings. Corroded chunks have been reported to detonate or ignite when cut with a knife.[87] Only fresh pieces should be used, and the metal protected from oxygen during purification. Other less reactive metals in wire or sheet form may be burnished or, in certain cases, treated chemically to remove corrosion or to activate by plating action (see the treatments of silicon,[88] zinc,[89] and magnesium[90]). Finally, bulk metal may be freshly milled, shaved, or turned to provide a fresh surface (aluminum, magnesium, copper).

Organometallics prepared separately or purchased from a supplier (Appendix II) must be examined for purity and, if in solution, assayed for active concentration. Solid organometallics should not only be checked against the physical properties reported in the literature (melting point, solubility, infrared and nmr spectra) but also tested for possible impurities. Thus not all commercial samples of diphenylmercury are equally good, but a chloride test is strongly positive for some. Both reactive solid and liquid organometallics can be hydrolyzed (Section D), and a spectral examination of the organic products can prove valuable. Liquids can be checked out by physical properties (boiling point and refractive index) and even by gas–liquid chromatography (glc) if suitable exit traps for poisonous and sensitive samples are devised. The quantitative and specific analysis of organometallic samples is a variegated and complex undertaking,[91] but certain commonly

---

[84] K. Ziegler, H.-G. Gellert, H. Lehmkuhl, W. Pfohl, and K. Zosel, *Justus Liebigs Ann. Chem.* **629,** 1 (1960).

[85] F. J. Radd and W. W. Woods, U.S. Patent 3,104,252 (September 17, 1963) [*Chem. Abstr.* **60,** 547 (1964)].

[86] C. W. Kamienski and D. L. Esmay, *J. Org. Chem.* **25,** 1807 (1960).

[87] W. S. Johnson and G. H. Daub, *in* "Organic Reactions" (R. Adams, ed.), Vol. VI, pp. 42–44, Wiley, New York, 1951.

[88] D. T. Hurd and E. G. Rochow, *J. Am. Chem. Soc.* **67,** 1057 (1945).

[89] C. R. Noller, *in* "Organic Syntheses" (A. H. Blatt, ed.), Coll. Vol. II, p. 184. Wiley, New York, 1943.

[90] H. Gilman and E. A. Zoellner, *J. Am. Chem. Soc.* **53,** 1581 (1931).

[91] T. R. Crompton, " Chemical Analysis of Organometallic Compounds," Vols. 1–5, Academic Press, New York, 1973–1977.

used schemes are worthy of mention. Methods employed for Grignard solutions and organolithium and organoaluminum reagents are discussed in Section D.

2. State of Subdivision

Since the reaction media of most organometallic reactions have low polarity, it is inevitable that metals and many salts will form a heterogeneous system. Particle size being critical to attaining feasible reaction rates, various methods of augmenting solid surface area have won favor: (*a*) machining of less active bulk metal to obtain chips or shavings, (*b*) cutting of a length of metal wire into small pieces with scissors and allowing the pieces to fall into a nitrogen-filled flask, (*c*) ball-milling of a metal or a salt in suspension in the presence of a chemical activator (e.g., ball-milling of aluminum powder in a suspension of triethylaluminum[92]), (*d*) spraying molten metal into the reaction vessel (the Ziegler process for aluminum alkyls),[84] (*e*) dispersing a molten metal in a hydrocarbon medium and then cooling (sodium dispersion in xylene[85,93]), (*f*) subjecting a metal sample to dispersion by ultrasonic waves (sodium in dioxane or in 1,2-dimethoxyethane[94]), (*g*) reduction of a metal salt in solution by chemical means (the formation of nickel from nickel acetylacetonate by aluminum alkyls[95]), and (*h*) the pyrolysis or chemical decomposition of metal complexes [the heating of dibenzenechromium(0)[96]]. As brought out in the foregoing section, trace impurities are influential, and therefore one should be sure not to contaminate the metal adversely when subdividing it.

3. Solvent

In addition to being properly purified, the ideal solvent for organometallic reactions has the following properties: (*a*) It is a good solvent for the starting materials and the desired organometallic product; (*b*) it promotes, or at least does not retard, the rate of the desired reaction; (*c*) it does not complex too tenaciously with the organometallic product, hence can be volatilized away; and (*d*) it is not attacked by the reagents or the products. Saturated

---

[92] K. Ziegler, H.-G. Gellert, H. Lehmkuhl, W. Pfohl, and K. Zosel, *Justus Liebigs Ann. Chem.* **629,** 12 (1960).

[93] See "Using Sodium in Dispersed Form," Ethyl Corporation pamphlet, Ethyl Corp., New York, New York.

[94] W. Slough and A. R. Ubbelohde, *J. Chem. Soc.* 918 (1957).

[95] K. Ziegler, H.-G. Gellert, E. Holzkamp, G. Wilke, F. W. Duck, and W.-R. Kroll, *Justus Liebigs Ann. Chem.* **629,** 188 (1960).

[96] B. D. Nash, T. T. Campbell, and F. E. Block, *U.S. Bur. Mines Rep. Invest.* 19 (1968), [*Chem. Abstr.* **69,** 38017 (1968)].

hydrocarbons come closest to being ideal solvents, but even these compounds have their drawbacks, such as low polarity and low solvating ability. Although organic halides, such as methylene chloride and 1,2-dichloroethane, may sometimes be suitable for less basic organometallics such as $RAlX_2$, the use of organic halides as solvents for organometallics of Groups IA–IIIA is not only inadvisable but also can lead to *explosions* (see refs. 64 and 108). Ethers, for example, ethyl ether, THF, and 1,2-dimethoxyethane, do not have these drawbacks; in fact, many organometallic reactions proceed at enhanced rates in such media. These reactions include the metalation of aromatic nuclei by alkyllithiums,[97] the addition of lithium alkyls to 1,1-diphenylethylene,[98] and the hydroboration of olefins.[99] On the other hand, ethers solvate the metal center firmly and are cleaved more or less readily by Group I metal alkyls and certain Group III alkyls and hydrides.[4,100] Hence the storage and isolation of certain metal alkyls in ether solution are curtailed, but such solutions can still be useful for immediate further reactions. In the event that the metal alkyl can be used as a solvate, an ether or tertiary amine medium may be suitable. The separation of aluminum alkyl etherates by distillation[101] and the isolation of Grignard compounds as amine complexes[102] exemplify this possibility.

For convenient reference, Table I lists the properties and peculiarities of frequently used solvents. Data on other suitable solvents are provided in Appendix III.

4. Mode of Admixing

To ensure prompt reaction, efficient agitation of the often heterogeneous mixtures must be maintained. Reactions conducted in completely sealed Schlenk tubes are best placed in a mechanical shaker. For routine reactions, the reactor may be equipped with a sealed blade stirrer (Fig. 19) or with a magnetic stirring bar. The Teflon coating may blacken during reactions with alkali metals, but no adverse effects seem to ensue. In reactions of dispersed alkali metals, a high-speed stirring arrangement (10,000 rpm) such as a magnetic Vibro-mixer employed with a creased or Morton flask (Fig. 21) may prove necessary.[103] Finally, a ball-mill reactor utilizing glass beads

---

[97] H. Gilman and S. Gray, *J. Org. Chem.* **23,** 1476 (1958).

[98] R. Waack, M. A. Doran, and P. E. Stevenson, *J. Organometal. Chem.* **3,** 381 (1965).

[99] See H. C. Brown, "Hydroboration," p. 4 Benjamin, New York, 1962, and references cited therein.

[100] L. I. Zakharkin *et al.*, *Izv. Akad. Nauk SSSR Otdel Khim. Nauk* 2255 (1959) [*Chem. Abstr.* **54,** 10837 (1960)].

[101] E. Krause and B. Wendt, *Chem. Ber.* **56,** 466 (1923).

[102] E. C. Ashby and R. Reed, *J. Org. Chem.* **31,** 971 (1966).

[103] Ace Glass, Inc., 1430 Northwest Boulevard, Vineland, New Jersey 08360.

TABLE I

SOLVENTS FREQUENTLY USED FOR ORGANOMETALLIC REACTIONS

Solvent	Formula	Boiling point (°C)	Melting point (°C)	Characteristics
Hydrocarbons				
Pentane	C_5H_{12}	36	−130	Generally suitable for dissolving or dispersing R_nM, but the arenes can be metalated by alkyls of Group IA or can form complexes with such Lewis acids [e.g., $(CH_3CH_2)_2Zn$]
Hexane	C_6H_{14}	69	−95	
Benzene	C_6H_6	80	−5.5	
Toluene	$C_6H_5CH_3$	111	−95	
Xylene (*meta*)	$C_6H_4(CH_3)_2$	139	−48	
Ethers				
Ethyl ether	$C_2H_5OC_2H_5$	35	−120	Form complexes of variable stability with R_nM of Groups IA–IIIA; undergo metallation and/or cleavage with variable rates; enhance rates of many nucleophilic (R^-) processes and retard rates of electrophilic (R_nM) processes.
Tetrahydrofuran	$(CH_2)_4O$	66	−65	
1,2-Dimethoxyethane	$CH_3OCH_2CH_2OCH_3$	85	−69	
Bis(2-methoxyethyl) ether	$(CH_3OCH_2CH_2)_2O$	162	−64	
Amines				
Pyridine	C_5H_5N	116	−42	Form firm complexes with R_nM of Groups IA–IIIA and change the reactivity markedly; pyridine can undergo addition reactions with most reactive M—R or M—H bonds
N-Methylpyrrolidine	$C_5H_{11}N$	81	−21	
N,*N*,*N'*,*N'*-tetramethylethylenediamine	$[(CH_3)_2NCH_2]_2$	122	−55	
Halides				
Methylene chloride	CH_2Cl_2	40	−97	Admixture with R_nM of Groups IA–IIIA can be *highly dangerous* (*possible explosion*)
Chloroform	$CHCl_3$	62	−64	
Phosphorus				
Hexamethylphosphoric triamide	$OP[N(CH_3)_2]_3$	232	7	Fosters carbanionic character of R_nM; promotes electron transfer pathways, as well; forms complexes and undergoes cleavage with R_nM

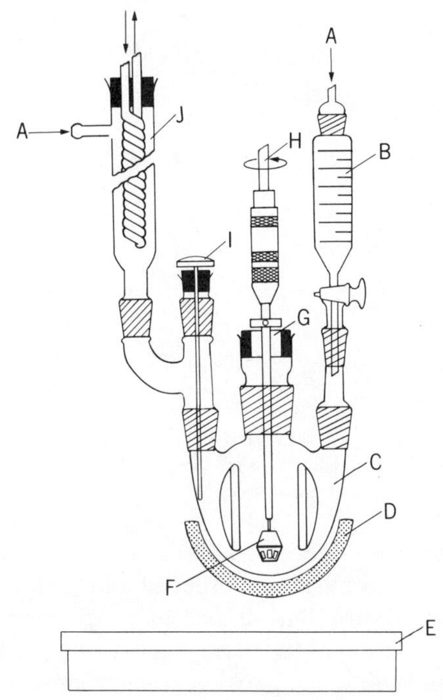

FIG. 21. Apparatus for high-speed stirring in a creased flask. A, Inert gas source; B, graduated addition funnel; C, creased flask; D, heating mantle; E, sheet iron safety pan; F, dispersator head, simplex or duplex model (Premier Mill Corp., Geneva, N.Y.); G, Oilite bearing fitted into rubber stopper (Amplex Division, Chrysler Corp., Detroit, Mich.). Immediately above Oilite bearing is a collar which prevents shaft from accidentally dropping through bottom of glass flask; H, stirrer, connected by means of flexible shaft to VB-2 grinding motor (R. G. Haskins Co., Chicago, Illinois), maximum speed, 18,000 rpm; I, thermometer (Model 2261, Weston Electrical Instrument Corp., Newark, New Jersey); J, reflux condenser fitted with a copper-tubing cooling coil; K, coolant. (Reprinted with permission from *Ind. Eng. Chem.* **45,** 404 (1953). Copyright 1953, American Chemical Society.)

(Fig. 22) has displayed high effectiveness in reactions of suspended transition-metal salts with Grignard reagents.[104] As a cautionary measure, a deep metal tray or bucket should surround any glass vessel whose contents are spontaneously ignitable.

The manner of admixing reactants with each other is especially important in many organometallic reactions where competing processes are dependent on the local concentration and ratio of reactants:

$$SiCl_4 \xrightarrow{C_6H_5Li} C_6H_5SiCl_3 \xrightarrow{C_6H_5Li} (C_6H_5)_2SiCl_2 \xrightarrow{C_6H_5Li} (C_6H_5)_3SiCl \qquad (8)$$

[104] Personal communication of Dr. W. Keim, 1966, then at the Max-Planck-Institut für Kohlenforschung, Mülheim (Ruhr), Federal Republic of Germany.

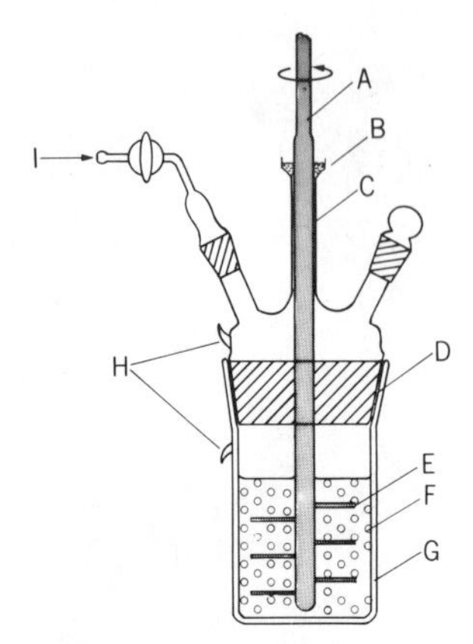

FIG. 22. Ball-mill reactor. A, Metal stirred shaft; B, lubricant reservoir; C, ground-glass bearing; D, ground joint; E, metal vanes on shaft; F, glass balls; G, heavy-walled glass reactor; H, fastners for metal spring; I, inert gas source.

Accordingly, besides the time of addition and the concentration of one reagent being added to another, the order of addition is crucial. The arbitrarily termed "normal addition" is one in which the substrate (namely, $SiCl_4$) is added to a solution of the more reactive organometallic. Obviously the local excess of phenyllithium favors oligophenylated silanes, even though the reactants are in an overall 1:1 ratio. Hence, to favor the formation of trichlorophenylsilane, adding the phenyllithium slowly to silicon tetrachloride in a so-called inverse addition is advised.

The technique of addition is essentially that employed in general preparative chemistry, with modifications dictated by the air and moisture sensitivity of the reacting system. For liquids, the pressure-equalizing dropping funnel shown in Fig. 6 of Volume 1 can be modified to advantage by inserting a stopcock in the connecting tube to prevent volatile contents from reaching the main vessel before desired and, conversely, hot vapors from condensing into the funnel. Alternatively, the equalizing connecting tube can be omitted and replaced by a gas lead near the top of the funnel. A further refinement is a platinum-wire needle valve that permits a steady drop rate.[105] In the case of solids, the funnel can be replaced by the adapted flask shown in Fig. 23.[106] If either liquids or solids need not be added in precise portions, some type of two- or three-chamber apparatus may suffice. This arrangement has the further merit that the components can be poured in together or can be

[105] Ace Glass Inc., 1430 Northwest Boulevard, Vineland, New Jersey 06380.

[106] The reservoir funnels of the Riley solids addition funnel apparatus offered by Ace Glass can be adapted to fit ground-glass inert gas connections and thus the funnels charged with one or two different solid reactants.

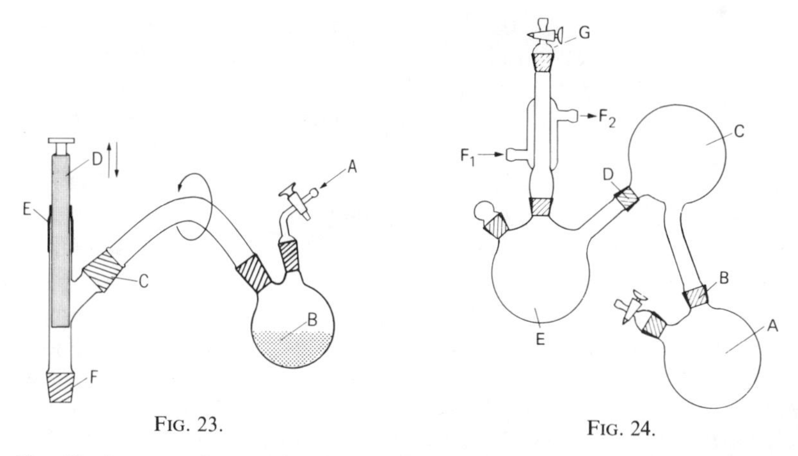

FIG. 23. FIG. 24.

FIG. 23. Apparatus for gradually introducing solids into a reaction mixture. A, Inert gas inlet; B, solid sample; C, rotatable Teflon joint; D, plunger for clearing solids out of tube F; E, snug rubber tubing for flexible, gastight seal.

FIG. 24. Dual- or triple-chamber vessel arrangements for reactions or recrystallizations (dual: A and C or, for reflux, C and E); F_1 and F_2, coolant flow; G, inert gas inlet.

volatilized over into the other by suitable reduction in pressure and heating or cooling. A simple type shown in Fig. 24 consists of two or three joined Schlenk flasks (A, C, and E). After the separate components are admixed by pouring from A to C, the solvent can be distilled from C back into A until crystals appear. Then the mother liquor is decanted from the crystals into A. Further recrystallization is effected by volatilizing the solvent back over onto the crystals and repeating the process. At the end the solvent and mother liquor in A can be cooled strongly and the connecting tube melted off. In a more flexible form, the vessels can be joined by ground joints (a modified Dreikugel-Apparatus), as shown in Fig. 24, where with care no freezing of joints nor undue contamination occurs. In other types of ground joint two-bulb chambers, a solid or liquid in a vessel (with or without a break tip (Fig. 31E) can be attached to a reaction vessel and turned to admix reagents. Finally, to introduce a gaseous reactant, one need only utilize the gas connections for the inert gas atmosphere (*caution:* do not contaminate the inert gas train) in order to pass gas over the surface of the reagent. If it is desirable to bubble gas through the liquid, a fritted gas inlet may be used. If a precipitate will result, however, a wide T-shaped gas inlet tube, with a glass rod acting as a plunger along the T-bar, will permit unclogging during reaction.

5. Temperature Control

As a supplement to the points raised on pp. 3–6 of Volume 1, the following comments are in order. Very low temperatures of -100 and $-78°$ are conveniently accessible by baths of liquid nitrogen, various solvents and solid carbon dioxide–ethanol in wide-mouth Dewar vessels, respectively.[107] The latter bath is superior to one using acetone, which tends to froth uncontrollably. An alcohol thermometer mounted in a neck of the reaction vessel proper may be desirable. For short-term cooling (e.g., the preparation of *n*-butyllithium in ether) a shallow earthenware crock or a styrofoam container set on a laboratory jack is most handy (*caution:* beware of breakage in raising the bath to the flask). Although solid carbon dioxide–ethanol, salt–ice, and ice water baths may often be used with no great hazard, such protic baths are most dangerous with violently hydrolyzable reaction mixtures in the event the flask should break. A safe alternative is a nonprotic, low-combustible bath (kerosene) which is circulated by a vortex or peristaltic pump through copper coils suitably cooled[108] (Fig. 25). The same precaution can be extended to the cooling of condensers or cold fingers of the reaction vessel (Volume 1, p. 5).[109]

For similar reasons of safety, hot water baths are unwelcome modes of heating. In addition, the condensation of water around joints of a scrupulously anhydrous reaction vessel can only be a source of anxiety and bother to the researcher. Oil baths equipped with some type of thermostatting (controlled by a hot-plate setting, an electronic contact thermometer, or a

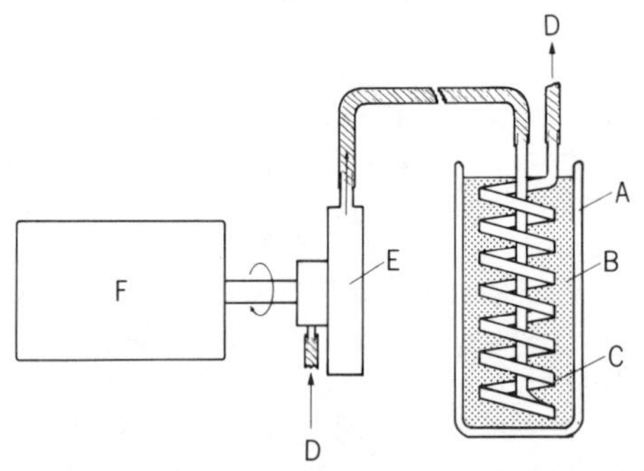

FIG. 25. Use of an inert coolant in large-scale organometallic reactions. A, Dewar container; B, appropriate cooling mixture; C, copper coils; D, flow of inert coolant through centrifugal pump E; F, motor.

---

[107] R. E. Randeau, *J. Chem. Eng. Data* **11**, 124 (1966).

[108] Although the use of halocarbons as cooling liquids might seem to avoid any chance of fire, such compounds react vigorously, and often explosively, with active metals and hydrides (ref. 64) and also with reactive organometallics (see H. Reinheckel, *Tetrahedron Lett.* 1939 (1964).

[109] As common practice in the Max-Planck-Institut für Kohlenforschung, Mülheim (Ruhr), Federal Republic of Germany, all-copper, cold-finger condensers were fashioned to fit into the ground joint of the elongated neck of a two-neck adapter (Fig. 21).

thermister)[110] will prove much more convenient. Alternatively, a refluxing solvent will make the use of an electric heating mantle both safe and feasible. By choice of condenser design (e.g., a Friedrichs-type condenser whose cold finger (Volume 1, Fig. 5) terminates only 2 cm above the bottom joint) and flask adapters, the ring of condensation in the condenser should be lower than the cooled stirrer bearing (Fig. 19). If this arrangement is not achieved, solvent loss, lubricant elution, and eventual sticking of the stirrer shaft will result. In certain cases involving two- or three-chamber reactions with higher-boiling solvents (bp $> 100°$), the upper part of the vessel may serve as a type of air condenser. A jet of compressed air will suffice for maintaining reflux.

6. Catalysts and Promoters

Additives that accelerate the reaction at hand, without being consumed (catalysts) or with sacrificial consumption (promoters), often are valuable in decreasing the time or increasing the yield of a synthesis. The exact basis of action of an additive may range among the following possibilities: (*a*) a scavenger for inhibiting residual amounts of water (MgX_2 in Grignard preparations),[111] (*b*) a more reactive chemical that attacks a metal surface readily and exposes a fresh surface (the use of ethyl bromide or ethylene bromide with bromopyridines and magnesium to promote the formation of pyridyl Grignards),[112] (*c*) a reagent that acts as a solubilizer of another reactant (the use of biphenyl for lithium metal reactions in THF),[113] (*d*) a metal that may establish a galvanic couple with a second metal (zinc–copper or copper–magnesium alloys),[114] or (*e*) a donor solvent whose solvation of metal ions speeds up alkali metal–organic substrate reactions (efficacy of crown ethers,[48] THF or glycerol ethers, in place of ethyl ether, for metal reactions).[115] The most general of possible experimental modifications involve varying the donor character of the solvent (Section B,3) or adding small amounts of a more reactive structural relative to promote a process. This latter process, termed "reaction by entrainment,"[112] can be used only

---

[110] Temperature controllers like the Thermistemp sold by Fisher Scientific Co. are suitable.

[111] See M. S. Kharasch and O. Reinmuth, "Grignard Reactions of Nonmetallic Substances," pp. 14–15. Prentice-Hall, Englewood Cliffs, New Jersey, 1954.

[112] (a) J. Overhoff and W. Proost, *Rec. Trav. Chim.* **57,** 179 (1938); (b) W. C. Davies and F. G. Mann, *J. Chem. Soc.* 276 (1944); (c) F. G. Mann and J. Watson, *J. Org. Chem.* **13,** 502 (1948).

[113] (a) J. J. Eisch, *J. Org. Chem.* **28,** 707 (1963); (b) J. J. Eisch and A. M. Jacobs, *J. Org. Chem.* **28,** 2145 (1963).

[114] H. Gilman and E. A. Zoellner, *J. Am. Chem. Soc.* **53,** 1581 (1931).

[115] (a) N. D. Scott, J. F. Walker, and V. L. Hansley, *J. Am. Chem. Soc.* **58,** 2442 (1936); (b) H. Normant, *Bull. Soc. Chim. Fr.* 728 (1957); (c) H. E. Ramsden, A. E. Balint, W. R. Whitford, J. J. Walburn, and R. Cserr, *J. Org. Chem.* **22,** 1202 (1957).

when the two resulting organometallics do not interfere with each other. Thus preparation of the 3-pyridyl Grignard reagent by the entrainment reaction of 3-bromopyridine and ethyl bromide with magnesium[112] is feasible, since in subsequent reactions (e.g., with allyl bromide) the 3-allylpyridine is easily separable from 1-pentene. Other common examples of entrainment are (*a*) the use of sodium amalgams to prepare mercury aryls from aryl halides.[116] (*b*) the use of lithium with 2% sodium for organolithium processes,[117] and (*c*) the use of ethyl bromide starter in ethyl chloride–aluminum metal processes.[118]

7. Precipitation and Filtration

Where precipitation occurs during reaction, provision for stirring by blade or paddle should be made. Especially heavy, sticky precipitates may require a twisted-wire (Hershberg) stirrer[119] to prevent local hot spots. Not only will hazardous bumping be avoided, but the resulting thorough mixing may promote the formation of a granular, more readily filterable precipitate. For granular, well-settling precipitates, either gravity filtration through the frit in the reaction vessel (Fig. 19) or decantation by siphon, pipet, or syringe (Figs. 13d and 14) is adequate. Addition of fresh, dry solvent to the residue, stirring, and similar removal of the wash solution complete the separation. In the event that the suspended solid is too gelatinous or too fine to settle well, the entire reaction mixture can be transferred to a centrifuge Schlenk tube (Fig. 13b) under nitrogen. Counterbalanced with a similarly filled Schlenk tube, the precipitate is spun down and the supernatant liquid siphoned or pipetted off.

Various adaptations for filtration by decantation through a glass frit have been described. A glass filter insert for two- or three-chamber vessels (Fig. 24) is shown in Fig. 26. If the occasion demands, such a filter can be fused into the connecting arm of the double Schlenk tube shown in Fig. 13d.

8. Gas Evolution

Permitting egress for a moderate evolution of gas during reaction is no problem, for the excess-pressure valve of the inert gas line can serve adequately. Should the evolution become too vigorous or frothing ensue, cooling the vessel and interrupting the stirring may keep it under control. Clearly,

---

[116] H. O. Calvery, *in* "Organic Syntheses" (H. Gilman and A. H. Blatt, ed.), Vol. 1, 2nd ed., p. 228. Wiley, New York, 1941.

[117] E. W. Kamienski and D. L. Esmay, *J. Org. Chem.* **25,** 1807 (1960).

[118] A. von Grosse and J. M. Mavity, *J. Org. Chem.* **5,** 106 (1940).

[119] E. B. Hershberg, *Ind. Eng. Chem. Anal. Ed.* **8,** 313 (1936).

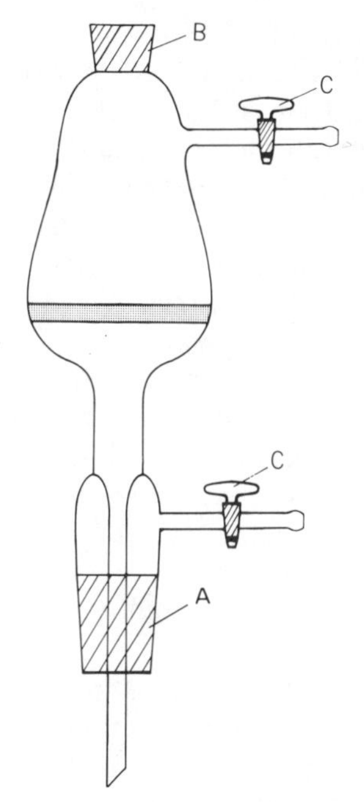

FIG. 26. Glass frit filtration unit. A, Drain joint; B, entry joint; C, inert gas entry (can be omitted if the frit is incorporated into the apparatus in Fig. 24 (at B or D).

provision for the safe disposal of flammable or noxious off-gases should be made.

Often the quantity of gas evolved is of interest in controlling or monitoring the extent of the reaction (e.g., the evolution of isobutylene in the preparation of tri-*n*-decylgallium from 1-decene and triisobutylgallium).[120] In such a situation, the mercury bubbler shown in Fig. 4 can be surmounted on the reflux condenser, the inert gas pinched off at F (Fig. 3), and the emitted gas discharged into a gasometer (Fig. 10). The mercury-type gasometer in Fig. 10 can be connected directly to the flask, while the large-volume salt solution type should have an intervening drying tube. The collected volumes are corrected for ambient temperature, pressure, and saline solution vapor pressure and may be sampled for mass spectrometric analysis.

9. Monitoring the Course of Reaction

As mentioned above, the quantity and the nature of the gas evolved can serve as a valuable check on the course of the reaction. Although admixed with air and inert gas, the collected gas can be analyzed for significant gas

[120] J. J. Eisch, *J. Am. Chem. Soc.* **84,** 3605 (1962).

ratios. Thus the thermolysis of neat triisobutylgallium yielded off-gas containing isobutylene and hydrogen in almost a 2:1 ratio, thereby revealing a breakdown via gallium hydride and isobutylene.[121] In a related manner, the gases arising from the hydrolysis of an organometallic reaction aliquot (see Section D,1 for the apparatus) can prove equally informative. For example, the treatment of triethylaluminum with gaseous hydrogen at 150° is allowed to proceed until a liquid aliquot yields almost a 2:1 ratio of ethane and hydrogen upon hydrolysis. Workup then gives an optimum yield of diethylaluminum hydride.[122]

A similar hydrolytic workup of reaction aliquots for liquids and solids now can be coupled with a host of monitoring techniques: glc, thin-layer chromatography, mass spectrometry, or some form of spectroscopy. The extent of halogen–lithium exchange with *n*-butyllithium and aryl bromides (arene/aryl bromide ratio on glc),[123] of hydralumination of alkynes (alkyne/alkene ratio),[124] or of alkyl group exchange in redistribution reactions of $R_mM-R'_mM'$ pairs (alkyl signals in the nmr spectra)[125] all can be determined readily.

In other situations, a labeling or derivatizing procedure with a series of standard aliquots may serve to mark the molecular site of the carbon–metal bond and to determine optimum reaction conditions.[126] Alternatively, samples may be titrated for active contents of carbon–metal bonds (Section D,2).

Two of the most common and useful tests for the presence of active carbon–metal bonds are Gilman's color tests I[127] and IIA.[128] The former can detect RM types capable of reacting with Michler's ketone:

$$R{-}M + \left[(CH_3)_2N{-}C_6H_4\right]_2C{=}O \rightarrow$$

$$\left[(CH_3)_2N{-}C_6H_4\right]_2C{<}^{OM}_{R} \xrightarrow[\text{2. HOAc, } I_2]{\text{1. } H_2O} \left[(CH_3)_2N{-}C_6H_4\right]_2C^{\oplus}{-}R \qquad (9)$$

Greenish blue

[121] J. J. Eisch, *J. Am. Chem. Soc.* **84,** 3830 (1962).

[122] Experiment of R. Köster described in Ref. 84, p. 13.

[123] H. J. S. Winkler and H. Winkler, *J. Am. Chem. Soc.* **88,** 964, 969 (1966).

[124] (a) J. J. Eisch and S. G. Rhee, *J. Am. Chem. Soc.* **96,** 7276 (1974); (b) J. J. Eisch and S. G. Rhee, *ibid.* **97,** 4673 (1975).

[125] K. Moedritzer, *in Adv. Organometal. Chem.* **6,** 171 (1968).

[126] See J. J. Eisch, "The Chemistry of Organometallic Compounds," Chapter 5, Macmillan, New York, 1967.

[127] H. Gilman and F. Schulze, *J. Am. Chem. Soc.* **47,** 2002 (1925).

[128] H. Gilman and J. Swiss, *J. Am. Chem. Soc.* **62,** 1847 (1940).

A positive response is typical of bonds between carbon and metals such as sodium, potassium, lithium, magnesium, calcium, barium, and strontium. A positive color test I naturally depends upon the other ligands in mixed alkyls R_nMZ_{m-n}, the solvent, and the concentration of reagents. The test has been modified by the use of other proton sources to detect arylmetallics in the presence of alkylmetallics.[129] Directions suitable for the execution of a general test are given below.

Color Test I. A 0.5 to 1.0-ml sample of the solution to be tested is treated, at room temperature, with an equal volume of a 1% solution of *N,N,N′,N′*-tetramethyl-*p,p′*-diaminobenzophenone (Michler's ketone) in dry benzene. After allowing the mixture to stand (up to 5 min for less reactive types), 1 ml of water is introduced slowly and with gentle shaking. Upon the subsequent addition of several drops of a 0.2% solution of iodine in glacial acetic acid, a characteristic greenish-blue color is formed if an active carbon–metal bond is present in the sample.

Color test IIA has an application limited principally to organolithium compounds, but the synthetic importance of these reagents makes such a test most helpful. In its most typical use color test IIA permits the detection of lithium alkyls in the presence of lithium aryls. The addition of *p*-bromo-*N,N*-dimethylaniline to an alkyllithium compound results in a lithium–bromine exchange to produce *p*-lithio-*N,N*-dimethylaniline. This compound then reacts with added benzophenone, and the resulting carbinol is converted to a red-colored carbonium ion by acid treatment:

$$R{-}CH_2Li \xrightarrow{Br{-}C_6H_4{-}N(CH_3)_2} (CH_3)_2N{-}C_6H_4{-}Li \xrightarrow{(C_6H_5)_2C{=}O} (CH_3)_2N{-}C_6H_4{-}C(OH)(C_6H_5)(C_6H_5) \longrightarrow (CH_3)_2N{-}C_6H_4{-}C^{\oplus}(C_6H_5)_2 \ \text{(Red)} \qquad (10)$$

[129] J. M. Gaidis, *J. Organometal. Chem.* **8,** 385 (1967).

Color Test IIA. A 0.5 to 1.0-ml sample of the organometallic solution to be tested is added to an equal volume of a 15% solution of *p*-bromo-*N*,*N*-dimethylaniline in dry benzene. Then 1.0 ml of a 15% solution of benzophenone in dry benzene is introduced. After a few seconds, the reaction mixture is treated with water and then acidified with concentrated hydrochloric acid. The appearance of a red color in the water signifies the presence of a carbon–metal bond able to undergo bromine–metal interconversion with *p*-bromo-*N*,*N*-dimethylaniline.

In order to monitor a reaction conveniently, be it for completeness of interaction or for the measurement of kinetics, one of the necks or orifices of the reaction vessel should be provided with a rubber septum (Fig. 13). Clearly, any needle syringe used for sample withdrawal should be gastight and scrupulously purged with an inert gas beforehand.

C. Purification of Organometallic Products

1. Consideration of Thermal and Light Instability

The degree of purity desired for an organometallic compound determines how elaborate a purification process will be necessary, or indeed if any isolation is necessary. Thus an X-ray crystallographic study of the unsolvated triphenylaluminum dimer[130] will dictate careful recrystallization from a hydrocarbon with the exclusion of any donor; the procurement of halogen- and ether-free *n*-butyllithium will lead to the choice of pentane for the interaction of lithium metal and *n*-butyl chloride and the subsequent separation of the lithium chloride[131]; but the formation of Grignard and organolithium reagents in an ether solution[132] can be followed by direct use without purification.

Attempted purification will not always succeed, because of the inherent instability of certain organometallic types and the close similarity in properties of isomeric mixtures of organometallics. The instability of organometallics can take various forms, among which are carbon–metal bond homolysis [Eq. (11)],[133] elimination [Eq. (12)],[124,134] cyclization [Eq. (13)],[135] disproportionation [Eq. (14)],[136] isomerization [Eq. (15)],[137]

---

[130] J. F. Malone and W. S. McDonald, *Chem. Commun.* 444 (1967).

[131] J. A. Beel, W. G. Koch, G. E. Tomasi, D. E. Hermansen, and P. Fleetwood, *J. Org. Chem.* **24,** 2036 (1959).

[132] (a) M. S. Kharasch and O. Reinmuth, "Grignard Reactions of Nonmetallic Substances," Prentice-Hall, Englewood Cliffs, New Jersey, 1954, pp. 1384; Houben-Weyl, "Methoden der Organischen Chemie," Vol. III/2a, p. 47. Georg Thieme Verlag, Stuttgart, Germany, 1973; (b) K. Ziegler and H. Colonius, *Justus Liebigs Ann. Chem.* **479,** 135 (1930); (c) For leading references on the preparation of organolithium reagents, see Houben-Weyl: "Methoden der Organischen Chemie," Vol. XIII/1, Georg Thieme Verlag, Stuttgart, Germany, 1970; B. J. Wakefield, "The Chemistry of Organolithium Compounds," Pergamon, Oxford, 1974.

[133] F. Glockling, *J. Chem. Soc.* 716 (1955).

[134] K. Ziegler, H. Martin, and F. Krupp, *Justus Liebigs Ann. Chem.* **629,** 14 (1960).

[135] J. J. Eisch and W. C. Kaska, *J. Am. Chem. Soc.* **84,** 1501 (1962).

[136] J. J. Eisch and W. C. Kaska, *J. Am. Chem. Soc.* **88,** 2976 (1966).

[137] P. T. Lansbury, V. A. Pattison, W. A. Clement, and J. D. Sidler, *J. Am. Chem. Soc.* **86,** 2247 (1965).

polymerization [Eq. (16)],[138] and insertion [Eq. (17)]:[139]

$$CH_2{=}\underset{\displaystyle CH_3}{C}{-}CH_2{-}Ag \xrightarrow{\Delta} Ag^0 + CH_2{=}\underset{\displaystyle CH_3}{C}{-}CH_2{-}CH_2{-}\underset{\displaystyle CH_3}{C}{=}CH_2 \quad (11)$$

$$(CH_3{-}\underset{\displaystyle CH_3}{CH}{-}CH_2)_3Al \longrightarrow (CH_3{-}\underset{\displaystyle CH_3}{CH}{-}CH_2)_2AlH + CH_2{=}\underset{\displaystyle CH_3}{C}{-}CH_3 \quad (12)$$

$$\text{2-biphenylyl-Al}(C_6H_5)_2 \longrightarrow \text{9-phenyl-9-aluminafluorene } (Al{-}C_6H_5) + C_6H_6 \quad (13)$$

$$(C_6H_5)_2AlH \xrightarrow{(C_2H_5)_2O} (C_6H_5)_3Al \longleftarrow O(C_2H_5)_2 + (AlH_3)_x \quad (14)$$

$$\text{cyclo-}(CH_2CH_2)C(H)CH_2Li \longrightarrow CH_2{=}CH{-}CH_2{-}CH_2Li \quad (15)$$

$$(CH_2{=}CH)_3Al \longrightarrow R{\left[CH_2{-}\overset{\displaystyle AlR_2}{CH}\right]}_x \quad (16)$$

$$CH_3{-}\underset{\displaystyle CH_3}{\overset{\displaystyle CH_3}{C}}{-}Li \xrightarrow[\text{from } (C_2H_5)_2O]{CH_2=CH_2} CH_3{-}\underset{\displaystyle CH_3}{\overset{\displaystyle CH_3}{C}}{-}CH_2CH_2Li \quad (17)$$

Even though all these modes of reaction are fostered by increased temperatures, they are also most sensitive to the presence of the donor solvent and certain catalysts.[140] For example, the disproportionation of unsymmetric alkyls ($R_nMR'_{m-n}$) and hydrides (R_nMH_{m-n}), as well as mixed alkyls (R_nMX_{n-m}), is favored by complexation of the amine[141] or ether[136] with the strongest Lewis acid member of the equilibrium[142]:

$$R_nMR'_{m-n} \rightleftharpoons R_mM + R_{m-1}MR' + \cdots + RMR'_{m-1} + R'_mM \quad (18)$$

On the other hand, the purification of certain organometallics as complexes $(C_6H_5)_2B{-}C{\equiv}C{-}Ph \cdot \text{pyridine})$[143] may prevent an undesired disproportionation.

Metal, either as a soluble zerovalent complex or as a heterogeneous phase, can accelerate the decomposition of many organometallic com-

[138] B. Bartocha, A. J. Bilbo, D. E. Bublitz, and M. Y. Grey, *Z. Naturforsch.* **16b,** 357 (1961).
[139] P. D. Bartlett, S. Friedman, and M. Stiles, *J. Am. Chem. Soc.* **75,** 1771 (1953).
[140] See J. J. Eisch, "The Chemistry of Organometallic Compounds," pp. 41–43, Macmillan, New York, 1967.
[141] E. Wiberg and T. Johannsen, *Naturwissenschaften* **29,** 320 (1941).
[142] A. W. Langer, Jr., *Ann. N.Y. Acad. Sci.* **295,** 110 (1977).
[143] J. Soulie and P. Cadiot, *Bull. Soc. Chim. Fr.* 1981 (1966).

pounds. Thus traces of nickel in the presence of 1-alkynes promote the elimination reaction shown in Eq. (12)[144]; in the preparation of bis-1,5-cyclooctadiene–nickel the formation of colloidal nickel speeds the deposition of the complex[145]; and the thermal decomposition of triisobutylgallium increases in rate with the deposition of metallic gallium.[120] Such autocatalytic decompositions mean that proper manipulations must minimize oxidations, thermolysis, and photolysis that generate traces of metals, and that rapid and efficient purification is often crucial for a successful preparation.

In addition to thermal instability, many organometallics exhibit moderate or pronounced sensitivity to light. Photodecomposition is especially noticeable in organometallics of heavier metals, such as those of cadmium, indium, lead, and mercury. Because of the autocatalytic decompositions noted above, storage and reaction vessels should be coated to exclude light where necessary. Moreover, when a previously stored sample is being used in a reaction or is being purified, *great caution should be exercised in any heating operation, lest thermal decomposition accelerated by a metal deposit lead to detonation.*

In the following sections various purification techniques will be discussed, with special consideration of the modifications and difficulties encountered with organometallics. Discussion of recrystallization, sublimation, and chromatography will be limited to special features only, since these techniques are dealt with in Volume 1.

2. Complex Formation

The intentional formation of complexes between organometallics and Lewis donors, such as halide ions, amines, ethers, and carbanions, can serve as a route to a stable organometallic derivative which may distill or recrystallize conveniently. After isolation and purification of the derivative has succeeded, the uncomplexed organometallic may be regenerated by a physical [Eq. (19)][146] or chemical process [Eq. (20)][147]:

$$AlCl_3 \text{ (in ether)} \xrightarrow{3C_6H_5Li} (C_6H_5)_3Al{:}O(C_2H_5)_2 \xrightarrow[-\text{pres.}]{200^\circ} (C_6H_5)_3Al \qquad (19)$$

$$BF_3 \text{ (in ether)} \xrightarrow[2.\ H_2O]{1.\ 4C_6H_5MgBr} (C_6H_5)_4B^- \xrightarrow{(CH_3)_3\overset{+}{N}H} (CH_3)_3\overset{+}{N}H\ \ \overset{-}{B}(C_6H_5)_4 \xrightarrow{\Delta} (C_6H_5)_3B \qquad (20)$$

[144] K. Ziegler, E. Holzkamp, H. Breil, and H. Martin, *Angew. Chem.* **67,** 541 (1955).
[145] Observations of J. E. Galle and K. R. Im of this laboratory.
[146] H. Gilman and K. E. Marple, *Rec. Trav. Chim.* **55,** 133 (1936)
[147] G. Wittig, G. Keicher, A. Rückert, and P. Raff, *Justus Liebigs Ann. Chem.* **563,** 118 (1949).

A second role in purification for such complex formation is to complex impurities, starting material, or by-products with a suitable Lewis base and then, by virtue of enhanced volatility or polarity differences, to distill off or to extract the desired product. The separation of the by-product, diethylaluminum chloride, from the product, triethylgallium, is effected by permitting potassium chloride to complex with the former and distilling off the latter[121]:

$$GaCl_3 \xrightarrow[2.\ 3KCl]{1.\ 3(C_2H_5)_3Al} (C_2H_5)_3Ga\uparrow + 3K[(C_2H_5)_2AlCl_2]\downarrow \quad (21)$$

The classic Schlenk precipitation of all magnesium halides and some alkylmagnesium halides from ether solution by dioxane[148] illustrates extraction from a complex:

$$2RMgX \rightleftharpoons R_2Mg + MgX_2 \xrightarrow{\text{dioxane}} R_2Mg + MgX_2(\text{dioxane})\downarrow \quad (22)$$

The factors governing the choice of a good complexing agent are implicit in the foregoing applications. For complexes with organometallic products, complex formation should not cause any irreversible chemical change in the product [Eqs. (11–18)]; the complex should be stable enough to survive the purification [Eqs. (9–10)]; the complex should be decomposable feasibly to yield the organometallic [Eqs. (19–20)]; and the donor should form a complex exclusively with the organometallic product. For complexes with by-products, the donor should instead complex exclusively with the impurity. The alkylation of gallium and indium halides with aluminum alkyls does not only pose the aforementioned need to separate dialkylaluminum chloride [Eq. (21)] but also to remove traces of any remaining aluminum alkyl. Addition of either sodium fluoride or potassium fluoride is effective, since indium alkyls complex with neither fluoride, gallium alkyls only with potassium fluoride, and aluminum alkyls with both fluorides.[121] A table delineating the ability of alkali metal halides to complex with aluminum alkyls has been devised by Ziegler and co-workers.[11]

The general experimental technique for complex formation introduces little new over the previous hints on filtering off solid phases or pipetting supernatant layers. Although complex formation of organometallics is often spontaneous with ethers and amines, gentle to prolonged heating with stirring may prove necessary with solid metal halides. Finally, coordination of certain organometallics (boron and aluminum aryls) with carbanions (lithium and magnesium organometallics) may be conducted titrimetrically, the end point being discerned by means of color test I[127] since $(C_6H_5)_3M$ and $(C_6H_5)_4M^-$ do not give a positive test.

---

[148] (a) W. Schlenk and W. Schlenk, Jr., *Chem. Ber.* **62,** 920 (1929); (b) W. Schlenk, Jr., *Chem. Ber.* **64,** 734 (1931); (c) G. O. Johnson and H. Adkins, *J. Am. Chem. Soc.* **54,** 1943 (1932).

3. Extraction

For organometallic (boron) and organometalloidal (tin, lead, etc.) compounds that are insensitive to protic sources, the initial separation of organic products and metal salt by-products can best be made by extraction with water. If the organometalloid warrants it, the hydrolysis might have to be conducted under an inert gas [tricoordinate (boron and arsenic)]. Whenever certain organometallics are especially sensitive to hydroxide ion [coordination with $(C_6H_5)_3B$[149] or silicon–hydride cleavage with $(C_6H_5)_3Si\text{-}H$],[150] the pH should be kept near 7 by the use of aqueous buffers.

With all other organometallics and with mixed organometalloids bearing hydrolyzable ligands [$(C_6H_5)_2PCl$ and $(C_6H_5)_2BCl$], a nonprotic extraction should be undertaken. As the most common situation for extraction involves Grignard or organolithium reactions in ether, the usual workup for nonvolatile organometallics is to remove the ether under reduced pressure. The desired organometallic is then extracted from the metal salt residue by means of dry benzene or petroleum ether, which may be warmed if necessary. To avoid ether altogether in the reaction, an organometallic of Group I or II may be prepared in a hydrocarbon suspension or solution, the reagent mixed with the substrate, and the solution filtered from inorganic salts to obtain the extract of the organometallic product.

The procedure for the extraction of water- and air-sensitive compounds can take a number of forms: (*a*) The extraction can be performed in the reaction vessel itself, especially one provided with a stopcock drain (Fig. 19); (*b*) separatory funnels[151] fitted with inert gas inlet tubes (Fig. 26 without frit) can be connected together (joint A of upper to joint B of lower) for the separation and preservation of both layers; (*c*) separate layers can be parted by pipetting or siphoning off either layer by adaptation of the technique described in Section A,5; and (*d*) organic extracts can be separated from a solid residue by filtration under an inert gas (Section B,7) or by means of a modified Soxhlet apparatus. In the dry box the solid to be extracted is loaded into a glass cup with a filter frit, and at the conclusion of the extraction flask A may be disconnected and hooked up to another flask for recrystallization.

4. Chromatography

Of the current modes of chromatography, gas–liquid, liquid–liquid, and adsorption, the first is the most readily adapted to organometallics, since the carrier gas (helium or nitrogen) furnishes an inert medium. By

---

[149] D. L. Fowler and C. A. Kraus, *J. Am. Chem. Soc.* **62,** 1143 (1940).

[150] L. Kaplan and K. E. Wilzbach, *J. Am. Chem. Soc.* **77,** 1297 (1955).

[151] See the "No-Air" apparatus offered by Ace Glass Company.

providing for the trapping of toxic, hydrolyzable, and oxidizable off-gases with an inert gas cold trap at the exit port, many pure organometallics can be collected for analytical purposes.[152] A suitable collection trap is illustrated in Fig. 27. The possibility of a thermal reaction on the column [Eqs. (11–17)] should not be ignored; some type of spectral monitoring before and after chromatography will reveal such undesirable thermal changes. The difficulty of finding suitably inert liquid phases and solid supports for glc severely limits its generality.

Although thin-layer absorption chromatography may be adapted for detecting the homogeneity of organometallics, preparative runs are most tedious. Column chromatography of air- and moisture-sensitive compounds may be undertaken, but the inertness of the organometallic to the eluting solvents and adsorbent must be ascertained beforehand. The ordinary chromatography column has to be equipped as shown in Fig. 28 for the introduction of solvent and the removal of eluate fractions under an inert atmosphere. Successful chromatographic separations require the scrupulous degassing and drying of all solvent and adsorbent.[153]

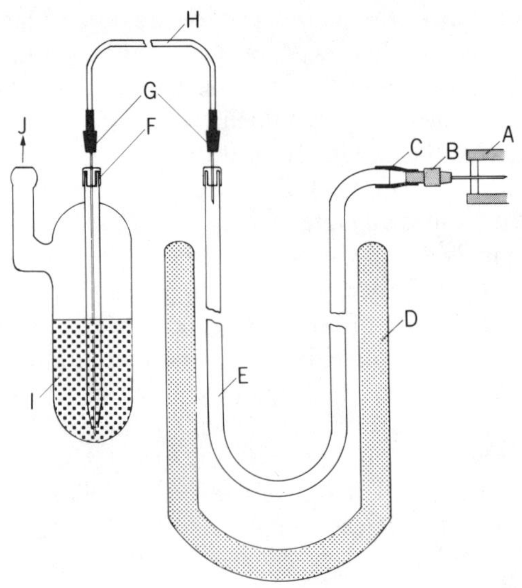

FIG. 27. Trap for collecting air-sensitive samples from the gas chromatograph. A, Gas chromatograph exit septum; B, steel needle and adapter; C, tight rubber sleeve; D, Dewar cooling bath; E, glass or stainless steel collecting tube; F, septa; G, double-needle Teflon (H) tubing; I, mineral oil bubbler; J, inert gas exit.

---

[152] See G. Schomburg, R. Köster, and D. Henneberg, *Z. Anal. Chem.* **170,** 285 (1959), for gas chromatographic analytical techniques suitable for boron alkyls.

[153] See D. A. Vyakhirev, V. T. Demarin, A. E. Ermoshin, and N. K. Rudnevskii, *Dokl. Akad. Nauk SSSR* **215,** 855 (1974) [*Chem. Abstr.* **81,** 20617 (1974)] for its use in separating bis-arenchromium iodides.

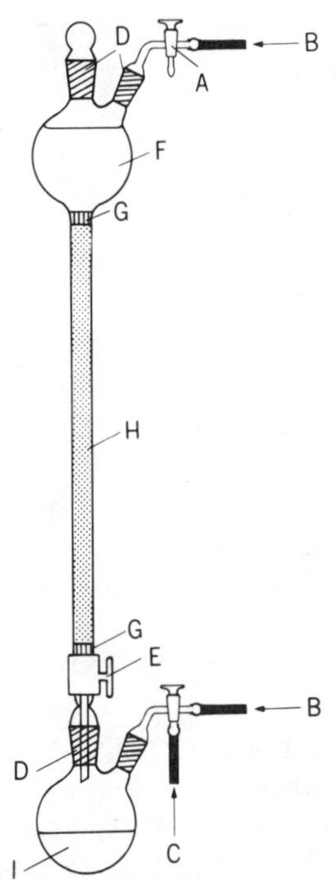

FIG. 28. Arrangement for the column chromatography of sensitive samples. A, Three-way stopcock for the alternate evacuation (C) and filling (B) of vessels with the inert gas; D, Teflon-coated joints; E, Teflon stopcock; F, eluting solvent; G, layer of sand; H, adsorbent; I, eluate.

Because of the enhanced possibility of thermal or chemical decomposition, column chromatography is usually a feasible purification only with less reactive organometallics (boron, silicon, tin, and transition-metal π complexes). In such cases, the further refinement of high-pressure liquid chromatography may enhance the efficiency of separation markedly, as in the preparation of substituted ferrocenes.[154]

5. Recrystallization

The contracted range of permissible solvents for crystallizing organometallics may occasion difficulties in procuring good recovery of well-formed crystals. Often dissolution in an aliphatic or aromatic hydrocarbon is achieved only by heating with large volumes of the solvent. To attain a hot saturated solution, which will not simply deposit ill-formed crystals too rapidly, is not an easy task. By recrystallizing the organometallic as a com-

[154] Personal communication from Dr. Manny Hillman, Brookhaven National Laboratory, Upton, New York 11973.

plex (etherate, aminate) not only may superior crystals be formed, but a wider range of permissible donor solvents may be utilized for recrystallization.

The recrystallization apparatus for the most exacting situation is the modified two- or three-chambered apparatus in Fig. 24, with a type of filter frit between vessels E and C. The filter frit can be wrapped with electrical heating tape for hot filtration. The crude solid is placed in E (the dry box) with the excess solvent, and the system is heated to dissolve all soluble material. After solution has been attained, E is tilted to let the solution flow through the filter frit into C. Under a nitrogen flow (from vessel A) vessel E and the frit are removed and vessel C stoppered at D. *Excess solvent is distilled from C into A (which is cooled). After crystallization in C is complete, the mother liquor is poured over into A.* The solvent is then distilled back from A to C, and the crystals redissolved hot in the fresh solvent. The procedure between the asterisks is repeated to effect the second recrystallization. Further recrystallizations can be repeated as desired. Vessel A is disconnected under streaming inert gas, and vessel C is capped at B and provided with a three-way stopcock at D. Vessel C is submitted to reduced pressure at D, with or without warming, to remove traces of solvent.

Often a cooled, supersaturated solution of an organometallic may be sluggish to crystallize. The impatient experimenter may wish to chill a metal rod in liquid air and apply it to the flask wall. The cold spot may then initiate crystal formation. In other cases, prolonged cooling at -30 or $0°$ in a Dewar bath may prove helpful.

Melting point capillaries (100 mm in length) can be filled conveniently in a dry box and temporarily stoppered with gum rubber stoppers bored only partially. Upon removal from the chamber the capillaries can be sealed with a microburner.

6. Distillation

The thermal destruction or chemical transformation of an organometallic reviewed in Section C,1 can become a serious problem in the process of distillation. For this reason, the volatilization is often advantageously conducted at as high a vacuum and over as short a period as possible. Special precautions should be taken to avoid hot spots, leaks, and contact with certain metal surfaces, which may unleash explosive decomposition. All this is by way of urging a well-controlled distillation for a safe and effective purification.

The distillation apparatus (Fig. 29) generally consists of a Schlenk vessel as the pot (K), a column and heat unit to avoid a contaminating joint (J), a fraction distributor with a drip tip (G) to deposit distillate directly into

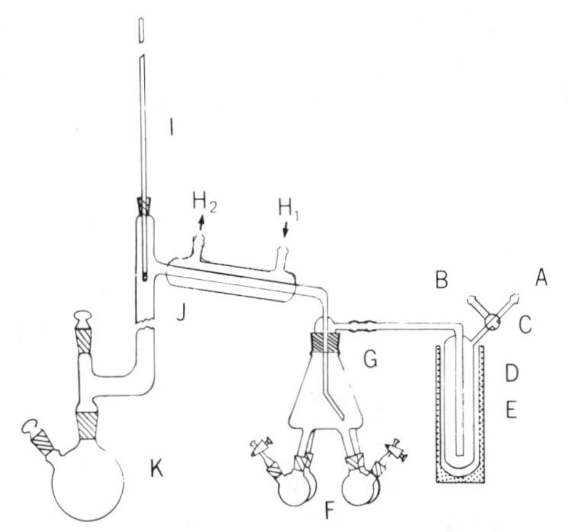

FIG. 29. Distillation apparatus for organometallic compounds. A, Vacuum source; B, inert gas; C, three-way stopcock; D, Dewar bath; E, cold trap; F, Schlenk vessel receivers; G, fraction distributor; H, inert condenser coolant (source: see Fig. 25); I, thermometer; J, side arm, which can bear a fractionation column; and K, Schlenk distillation flask equipped with Teflon-coated joints.

receivers without joint contamination, graduated Schlenk tube receivers (F), cold traps (E) to protect the oil pump or to prevent back-diffusion of moisture with water aspirators, condenser coolant (H) of a safe, inert variety (Fig. 25), a small bar magnet stirrer for the pot contents, and/or a capillary bleeder for the inert gas.[155] Furthermore, the pot should be heated with an oil bath equipped with thermal regulation. For the eventuality of a sudden temperature rise, it is advantageous to maintain the heating bath on a laboratory jack for quick removal. In addition, the pressure at which the distillation is conducted may feasibly be somewhat higher than that which the pump can attain. Suitable manostats for distillation in the pressure range of 5–760 mm can be devised or purchased and then suitably adapted for distilling organometallics.[156]

For cases of distilling high-boiling liquids and low-melting solids from very volatile impurities, various styles of short-path, wide-bore apparatus have been proposed. Typical of these is an apparatus devised by Wittig and co-workers[147] for the preparation and distillation of triphenylborane [Eq. (20)], whose provision with capillaries permits the distillate to be transferred to ampuls by the technique described in Section A,5,a.

[155] A variety of specialized distillation setups for organometallics can be found in H. Metzger and E. Muller, Houben-Weyls "Methoden der Organischen Chemie," Vol. I/2, p. 351. Georg Thieme Verlag, Stuttgart, Germany, 1959.

[156] Vacuum regulators permitting organometallic vapors to come into contact with mercury or metal wire are unsuitable; an all-glass apparatus of the Cartesian diver type is superior.

The performance of a distillation begins by a complete setting-up and flushing of the apparatus with inert gas. The apparatus should be checked for maintenance of high vacuum during this process. The liquid to be distilled, previously freed of most solvent, is transferred to the pot (see Section A,5 for technique). Without heating, the balance of the solvent is removed and caught in the solid carbon dioxide–acetone trap. The trap is cleaned out and reflushed before final distillation of the pot residue. Upon completion of distillation and cooling, inert gas is admitted at A and the ebullator is removed and replaced by a three-way stopcock. In this manner the contents of K and E can be saved.

Safety precautions during distillation should assume anything could go wrong. The whole setup is placed in a 2-in.-deep metal trough and isolated from the operator by stable safety shielding. No water should be used either as coolant in condensers or as a heating bath. With water aspirators a capacious safety flask should be used at the pump with a shutoff valve for water flowback. A drying tower (P_2O_5) or a liquid air trap should complete the pump train.

D. Characterization of Organometallic Compounds

1. Physical Properties

The measurement of physical constants for many organometallics requires rather elaborate experimental procedures. Although there are superior analytical and spectroscopic methods for identifying isolated organometallics, the invariability of physical data on purification is a strong proof of homogeneity and, usually, purity. The quantities melting point, boiling point, refractive index, density, and dipole moment are those of most general utility. Some type of inert gas chamber is indispensable either for transfer to a measuring tube or for the actual conducting of the measurement. Consider the description in Section C,5 for the filling of melting point capillaries, on the one hand, and the placement of an Abbe refractometer and organometallic sample in a collapsible polyethylene bag,[35] on the other. For the latter situation, the bag can be evacuated flush against its contents and then refilled. After the purging is complete, the organometallic is placed on the prism by manipulating through the bag wall, and the measurement is made. As illustrated below, variants comprise the kinds of cells and vessels that can be filled under an inert gas. The melting points or boiling points of organometallics can be determined in the usual way, in sealed capillaries or in the course of a distillation, respectively. Because of the possibility of thermal decomposition, a melted sample should be cooled and remelted, and a liquid sample redistilled, before placing complete confidence in the first observation. Other signs of chemical change, such as color change, frothing, resolidifying, and remelting, should be duly noted. Associated liquids, for example, $[(C_2H_5)_2AlF]_x$, do not actually distill but undergo depolymerization and evaporative transport upon heating.[157] Likewise, triphenylaluminum etherate melts at 128–129° but resolidifies toward 200° as ether is released from the complex, and finally melts at over 200°[158] as triphenylaluminum (lit. mp 242°).[159] For thermally labile compounds,

---

[157] K. Ziegler and R. Köster, *Justus Liebigs Ann. Chem.* **608,** 1 (1957).

[158] G. Wittig and D. Wittenberg, *Justus Liebigs Ann. Chem.* **606,** 13 (1957).

[159] (a) A. N. Nesmeyanov and N. N. Novikova, *Bull. Acad. Sci. USSR* 372 (1942); (b) E. Krause and B. Wendt, *Chem. Ber.* **63,** 2401 (1930).

melting points may be achieved with carefully preheated baths, while the boiling point measurement may have to be replaced by a measurement of vapor pressure as a function of temperature in a tensimeter. By graphing the logarithm of vapor pressure versus temperature, the value of the theoretical boiling point can be estimated.[160]

The density and refractive index, necessary data for computing molecular refraction, do not have the same prominence nowadays as they enjoyed 30 years ago. Other more fundamental measurements are now available. Although some estimate of density is handy for arriving at reaction proportions

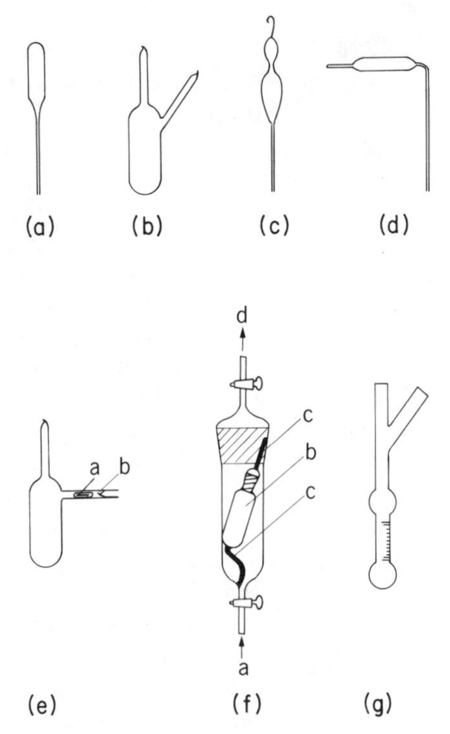

FIG. 30. Ampuls and pycnometers for solid or liquid organometallic samples. (a) Simple liquid-sample ampul (see Fig. 20); (b) sealed Schlenk tube (amber glass for photosensitive samples; see Fig. 17); (c) ampul provided with a glass hook (suitable for apparatus in Fig. 20); (d) ampul with right-angle entry tube; (e) sealed tube with magnetic-actuated (a) break tip (b); (f) miniature inert gas chamber whereby samples can be loaded from vials suspended in the current of inert gas (a → d) into vessels (b) provided with glass rod supports (c) and a ground stopper (large top joint is removed during transfers, and by use of forceps and solids tube (Fig. 15) the transfer is completed); and (g) Schlenk tube calibrated to serve as a pycnometer.

---

[160] H. Kienitz, *in* Houben-Weyls, "Methoden der Organischen Chemie," Vol. III/1, p. 295. Georg Thieme Verlag, Stuttgart, Germany, 1955.

by volume alone, the refractive index of sensitive liquids is hard to measure routinely and possibly not wholly necessary. By using calibrated and tared Schlenk tube receivers for the distillation of liquids (G in Fig. 30), densities of up to three significant figures can readily be obtained. Smaller quantities of liquids can be handled with pycnometers[161,162] that permit densities of four figures to be obtained. The latter type of course must be filled and sealed in a dry box or a Ziegler trough. For refractive index, the collapsible glove box discussed above can be supplanted with a type of Pulfrich[163] chamber, filled in a dry box, and measured in the usual way outside the box.

Other measurements of occasional interest are those for dipole moment,[164] conductivity,[165] crystal structure,[166] heats of combustion,[167] heats of solution,[168] and heats of association.[168,169]

2. Elemental Analysis

The scope and limitations of microanalysis for nonmetallic elements in organometallic compounds need be given no special treatment here (but see ref. 91). The principal problems reside in developing suitable methods for taring samples without oxidation and in preventing combustion train problems such as too violent an initial combustion or the retention of carbon as a metal carbide. Preliminary consultation with an experienced microanalyst[170] suggests that previously tared samples be submitted in microampuls for liquids or "piggies" for solids [(a)–(f) in Fig. 30].

---

[161] F. Hein, E. Petzchner, K. Wagler, and F. A. Segitz, *Z. Anorg. Chem.* **141,** 195 (1924).

[162] K. Ziegler, H. G. Gellert, H. Martin, K. Nagel, and J. Schneider, *Justus Liebigs Ann. Chem.* **589,** 119 (1954).

[163] E. Krause and A. von Grosse, "Die Chemie der Metallorganischen Verbindungen," p. 807. Verlag Gebr. Borntraeger, Berlin, 1937.

[164] See E. G. Hoffmann and G. Schomburg, *Z. Elektro. Chem.* **6,** 1101, 1106 (1967) and E. G. Hoffmann and W. Tornau, *Z. Anal. Chem.* **186,** 231 (1962) for their use of dielectric constant measurements in determining donor complex formation with organoaluminum compounds.

[165] See W. Strohmeier, H. Landsfeld, and F. Gernert, *Z. Elektro. Chem.* **66,** 823 (1962), for measurements on organolithium, -sodium, and -potassium compounds in various ethers.

[166] See H. Dietrich, *Acta Crystallogr.* **16,** 681 (1963) and E. Weiss and co-workers, *J. Organometal. Chem.* **2,** 197 (1964) and **21,** 265 (1970) for X-ray studies on tetrameric methyl- and ethyllithiums.

[167] See S. Pawlenko, *Chem. Ber.* **100,** 3591 (1967); **102,** 1937 (1969), for the apparatus and procedure as applied to organoaluminum compounds.

[168] See M. B. Smith, *J. Phys. Chem.* **71,** 364 (1967), and succeeding papers; *J. Organometal. Chem.* **22,** 723 (1970); **46,** 31 (1972); **46,** 211 (1972); **70,** 13 (1974) for measurements with organoaluminum compounds.

[169] See J. Brandt and E. G. Hoffmann, *Brennstoff Chem.* **45,** 200 (1964), for heats of mixing obtained from pairs of aluminum alkyls.

[170] See O. Schwarzkopf and F. Schwarzkopf, *in* "Characterization of Organometallic Compounds" (M. Tsutsui, ed.), in the series, "Chemical Analysis" (P. J. Elving and I. M. Kolthoff, eds.), Vol. 26, p. 35. Wiley (Interscience), New York, 1969.

The determination of the percentage of metal in a sample may be very simply achieved or require elaborate treatment, depending upon the volatility, the instability, the specific metal, and the complexity of the organometallic. With the simplest method, one contrives to weigh an analytical sample of the organometallic in a Vycor or porcelain crucible or boat without weight change due to oxidation, hydrolysis, or volatility. The sample is hydrolyzed and/or oxidized in a hydrocarbon solution, the solvent volatilized, and the residual cautiously treated with concentrated nitric or sulfuric acid. By successively gentle to strong heating the metal is finally weighed as its oxide. The procedure detailed below describes the analysis for gallium in rather volatile and reactive gallium alkyls.[120]

If the organogallium compound is a liquid, a 0.1 to 0.2-g sample is introduced into a pretared glass ampul (Figs. 20 and 30). The ampul is placed in a tared porcelain or Vycor glass crucible, and 1–2 ml of anhydrous ether introduced. The ampul is crushed carefully with the transversely flattened end of a glass stirring rod. The resulting solution is treated with 6 *N* sulfuric acid droplets in a gradual manner.

With solid organometallic samples, a tared crucible is charged with the sample in the inert atmosphere chamber, and the crucible is placed in a small, tared, screw-cap metal container.[171] The sealed container is removed from the chamber and weighed. Then the charged crucible can be taken out of the metal container, the ether diluent added as above, and the treatment with 6 *N* sulfuric acid undertaken.

The ether is evaporated from the hydrolyzed sample, and the crucible residue covered with 5–10 drops of concentrated sulfuric acid. The crucible is positioned in the Nichrome-wire triangle of a ring clamp, and a ring burner is placed so as to warm the uppermost edge of the crucible. By gentle heating in a hood the gallium sample is decomposed and the sulfuric acid evaporated as the ring flame is gradually brought to bear lower on the crucible. When dry, the sample residue is ignited to constant weight by means of a Mekler burner. Because of its hygroscopic nature, the gallium(III) oxide sample is allowed to cool in a dessiccator and the covered crucible is weighed rapidly. The precision of duplicate analyses is well within $\pm 0.30\%$, and the ampul fragments do not interfere with accurate analysis.

The rather generous sample size is occasioned by the errors due to decomposition and losses during hydrolysis. With more stable organometalloids the oxidation with sulfuric acid can be conducted with only milligrams in a platinum combustion boat placed in a 12-mm tube wound with no. 60 Nichrome wire and powered by a variable transformer.

If the compound is too volatile or reactive, or contains another metal or

[171] Procedure of C. K. Hordis, The Catholic University of America, Washington, D.C., 1966.

TABLE II

SPECIFIC ANALYTICAL PROCEDURES FOR METALS AND LIGANDS IN AN ARRAY OF MAIN GROUP ORGANOMETALLICS

Compound	Atom or group measured	Method	Reference
Group IA alkyls	C, H, Li, Na, or K	Combustion	*a*
$CH_2{=}CHLi$	C, H	Combustion	*b*
$KB(C_6H_5)_4$	K	Flame photometry	*c*
$(CH_3)_2Be$, R_2Mg	C, H	Combustion	*d*
R_2Be, R_2Zn, R_3B, R_3Al	R_nM	Gas chromatography	*e*
R_2Zn	Zn	EDTA titration	*f*
R_2Hg	C, H, Hg	Combustion	*g*
RHgX	Hg	Combustion (no loss of HgX_2)	*h*
R_3B	C, H, B	Combustion–titration	*i*
R_3Al	Al	EDTA titration	*j*
R_3Ga	Ga	H_2SO_4 decomposition and ashing	*k*
R_3In	In	H_2SO_4 decomposition and ashing	*k*
$C_6H_5TlCl_2$	$RTiCl_2$	Compleximetric titration	*l*
R_4Si	Si	Na_2O_2 fusion and SiF_6^{2-} titration	*m*
R_4Ge	Ge	Spark spectroscopy	*n*
n-$C_4H_9SnCl_3$	$RSnCl_3$	Complexation and EDTA titration	*o*
R_4Pb	Pb	H_2SO_4 decomposition and EDTA titration	*p*
$R_nAl(OR)_{3-n}$	OR	Hydrolysis and spectroscopy of ROH	*q*
R_nAlH_{3-n}	Al—H	Protolysis with $C_6H_5NHCH_3$ and gasometric analysis	*r*
RMgX, R_nAlX_{3-n}, R_nSiX_{4-n}, etc.	X	Hydrolysis and Volhard method for X	—

[a] A. S. Zabrodina and U. P. Miroshina, *Vestn. Moskov Univ.* **2,** 195 (1957).
[b] E. C. Juenge and D. Seyferth, *J. Org. Chem.* **26,** 563 (1961).
[c] M. G. Reed and A. D. Scott, *Anal. Chem.* **133,** 773 (1961).
[d] E. L. Head and E. Holley, *Anal. Chem.* **28,** 1172 (1956).
[e] P. Longi and R. Mazzocchi, *Chim. Ind. Appl.* **48,** 718 (1966).
[f] T. R. Crompton, "Chemical Analysis of Organometallic Compounds," Vol. V, p. 378, Academic Press, New York, 1977.
[g] T. Mitsui, K. Yoshikawa, and Y. Sakai, *Microchem. J.* **7,** 160 (1963).
[h] T. Sudo and D. Shinoe, *Jpn. Anal.* **4,** 88 (1955).
[i] R. C. Rittner and R. Culmo, *Anal. Chem.* **34,** 673 (1962).
[j] Reference *f*, p. 115.
[k] J. J. Eisch, *J. Am. Chem. Soc.* **84,** 3605 (1962).
[l] B. L. Pepe and R. Rivarola-Barbieri, *Anal. Chem.* **40,** 432 (1968).
[m] A. Sykes, *Microchim. Acta* 1155 (1956).
[n] R. P. Kreshkov and E. A. Kucharev, *Zavod. Lab.* **32,** 558 (1966).
[o] J. Efer, D. Quaas, and W. Spichale, *Z. Chem. Leipzig* **5,** 390 (1965).
[p] L. C. Willemsens and G. J. M. van der Kerk, "Investigations in the Fields of Organolead Chemistry," p. 87, International Lead-Zinc Research Organization, New York, 1965.
[q] T. R. Crompton, *Analyst* **86,** 652 (1961
[r] W. P. Neumann, *Justus Liebigs Ann. Chem.* **629,** 23, 29 (1960).

halogen, the decomposition should be conducted by opening a tared vial under an organic solvent and slowly adding a hydrolyzing reagent. The organic solvent can then be evaporated and the desired metal precipitated as a hydrous oxide or as a chelate such as oxinate (8-hydroxyquinoline) or dimethyloximate.[172] Filtration and drying of the chelate or ignition to the oxide can then be carried out. As an alternative to such gravimetric methods, volumetric titration with ethylenediaminetetraacetic acid can be employed (Volume 1, p. 50).

For orientation when consulting the copious literature on such analyses, we have compiled in Table II references to detailed descriptions of successful analytical procedures for individual metals in various organometallics. Also included are analytical techniques for commonly encountered ligands, such as solvate (R_2O, R_3N), halide, hydride, organic acid group, alkoxide, and metal–metal bonds.

3. Molecular Weight

The determination of molecular weight is especially crucial for organometallics of the main group or organometalloids, where oligomers or polymers of the empirical formula frequently are encountered. Although estimates of the molecular unit can be obtained by ebullioscopic, cryoscopic, osmometric, mass spectrometric, and X-ray crystallographic methods, no one technique suffices.[173] First, states of association of empirical formula units ($[R_mM]_x$) vary from the solid state to that in solution. Second, the value of x can vary not only from one solvent to the next, but even in a given solvent more than one oligometic species ($x \neq x'$) can be present. The complicated and not completely clear molecularity of organolithium compounds in solutions points up this situation.[174] The confusion and variety of results with Grignard reagents are further instances.[175] Third, it therefore follows that the specific nature of R, the concentration, and the temperature of measurement are extremely important in determining molecular weights in a given solvent.

---

[172] See reference 170, pp. 63–69.

[173] J. J. Eisch, "The Chemistry of Organometallic Compounds," Chapter 3. Macmillan, New York, 1967.

[174] See the discussion in Houben-Weyls, "Methoden der Organischen Chemie," and in Wakefield, as cited in ref. 132c.

[175] See K. Nützel, in Houben-Weyls "Methoden der Organischen Chemie," Vol. XIII/2a, p. 508. Georg Thieme Verlag, Stuttgart, Germany, 1973.

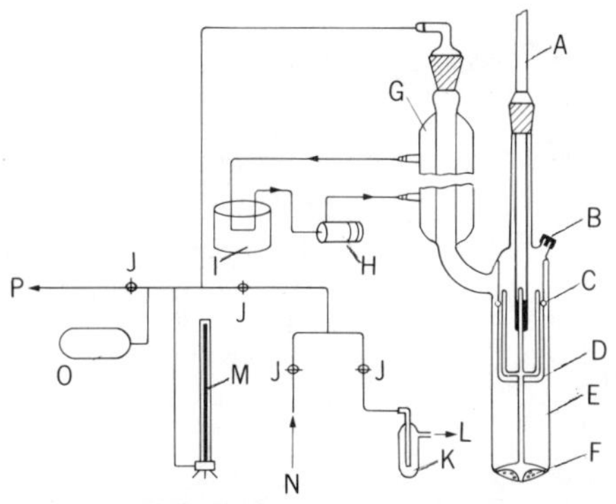

FIG. 31. Apparatus for the ebullioscopic determination of the molecular weight of air-sensitive compounds. A, Beckmann thermometer; B, septum for introducing sample; C, vapor outlet holes; D, thermometer shield; E, Cottrell pump; F, fused glass chips; G, condenser jacket with circulating pump H and temperature bath I; J, valves; K, gas bubbler; L, excess gas exit; M, manometer; N, nitrogen inlet; O, surge tank; P, vacuum. (Reprinted with permission from F. W. Walker and E. C. Ashby, *J. Chem. Educ.* **45,** 654 (1968). Copyright, 1968.)

The apparatuses useful for measuring molecular weights by boiling point elevation[176,177] and freezing point depression are shown in Figs. 31 and 32, respectively.[178] The more recently available vapor-phase osmometric technique,[179] allowing the rapid measurement of molecular weights by extrapolation to infinite dilution, seems destined to yield reliable data for various concentrations. In the course of determining the correct structure for an oligomer of $(C_6H_5)_2Si$ units, osmometric measurements of the molecular weight were found to be more reliable than cryoscopic determinations.[180]

The mass spectrometric measurement of molecular weights is most applicable when a solvent-free, homogeneous oligomer is under study. Provided the oligomer is not too unstable with respect to smaller units and the carbon–metal bonds are not completely disrupted at the instrument's operating voltage, the ions derived from the oligomer may be observable. Thus the mass spectrum of dimeric dimethylaluminum chloride shows

[176] G. Wittig, F. J. Meyer, and G. Lange, *Justus Liebigs Ann. Chem.* **571,** 199 (1951).

[177] F. W. Walker and E. C. Ashby, *J. Chem. Ed.* **45,** 654 (1968).

[178] T. L. Brown, R. L. Gerteis, D. A. Bafus, and J. A. Ladd, *J. Am. Chem. Soc.* **86,** 2135 (1964).

[179] A vapor pressure osmometer is available from Mechrolab, Inc., Mountain View, California.

[180] H. Gilman and G. L. Schwebke, *J. Am. Chem. Soc.* **85,** 1016 (1963).

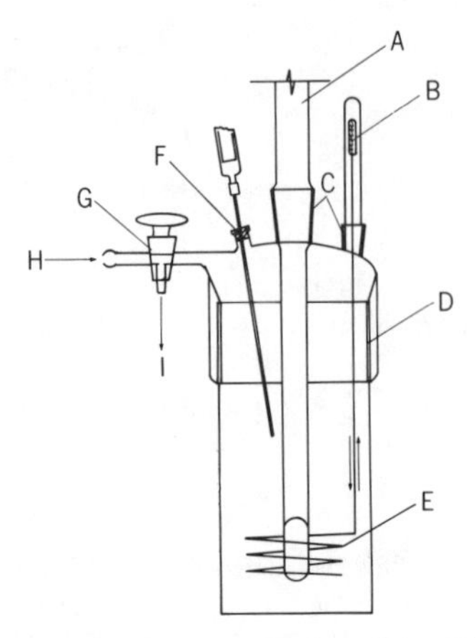

FIG. 32. Apparatus for the cryoscopic determination of the molecular weight of air-sensitive compounds. A, Beckmann thermometer; B, magnet, externally moved vertically to raise and lower stirring helix E; C, ground joints; D, large ground joint for attaching head of apparatus; F, septum for introducing sample; G, three-way stopcock for inert gas (H) or for vacuum (I). (Reproduced with permission from the Ph.D. Thesis of J. R. Sanders, School of Chemistry, Georgia Institute of Technology, 1968, pp. 48–50; we are most grateful to Professor E. C. Ashby for this assistance.)

$Me_3Al_2Cl_3^+$ with an intensity of 48.4 relative to the base peak of Me_2Al^+; and dimethylaluminum hydride shows $Me_3Al_2H_2^+$ with an intensity of 39.6 relative to Me_2Al^+.[181] The mass spectrum of ethyllithium provides evidence for the presence of tetrameric and hexameric aggregates.[182]

Lowering the sample pressure in the ionization chamber and the operating potential (~10 eV) may give assurance of the peak's identity. With organometallics of metals occurring as isotopes, each isotope should yield a parent peak with an intensity equal to the isotope's natural abundance. Thus triphenylborane yields parent peaks at 241 and 242 in a ratio of 1:4.

In Table III are listed reported, and sometimes conflicting, molecular weight estimates for organometallics obtained by a variety of methods. It is noteworthy that solution techniques apparently are frustrated in many cases by failure to find a solvent in which the solute exhibits ideal-solution behavior. X-ray crystallographic techniques, giving the dimensions of the unit cell and the density, permit a molecular weight to be assigned for otherwise troublesome compounds (e.g., the oligomers of diphenylsilanes).[180]

---

[181] J. Tanaka and S. R. Smith, *Inorg. Chem.* **8,** 265 (1969).

[182] J. Berkowitz, D. A. Bafus, and T. L. Brown, *J. Phys. Chem.* **65,** 1380 (1961).

TABLE III

EXPERIMENTAL VALUES FOR THE MOLECULAR WEIGHTS OF VARIOUS MAIN GROUP ORGANOMETALLIC COMPOUNDS

Compound	Medium	Association factor X in $[R\ M]_x$	Method	Reference
CH_3CH_2Li	Vapor	4 and 6	Mass spectrometry	*a*
$(C_6H_5)_3CLi \cdot TMEDA$	Solid	1	X-ray analysis	*b*
$(CH_3)_2Be$	Solid	Polymer	X-ray analysis	*c*
$(CH_3CH_2)_2Be$	Benzene	2	Cryoscopy	*d*
$(tert\text{-}C_4H_9)_2Be$	Vapor	1	Electron diffraction	*e*
RMgCl	Ethyl ether	2	Ebullioscopy	*f*
RMgBr, RMgI	Ethyl ether	1 (0.1 mole/liter), 2 (0.3 mole/liter)	Ebullioscopy	*f*
RMgX	Tetrahydrofuran	1	Ebullioscopy	*f*
$C_6H_5MgBr \cdot 2(CH_3CH_2)_2O$	Solid	1	X-ray analysis	*g*
$(CH_3)_3Al$	Solid	2	X-ray analysis	*h*
$(CH_3)_3Al$	Vapor	1 and 2	Electron diffraction	*i*
$(CH_3)_3Ga$	Benzene	1	Vapor pressure osmometry	*j*
$(CH_2{=}CH)_3Ga$	Cyclohexane	2	Cryoscopy	*k*
$(CH_3)_3In$	Benzene	4	Cryoscopy	*l*
	Benzene	1	Vapor pressure osmometry	*m*
$(C_6H_5)_2Si$	Benzene	4 and 5	Ebullioscopy, X-ray analysis	*n*
$(CH_3)_3Sn$	Benzene	1 and 2	Cryoscopy	*o*
	Benzene	2	Ebullioscopy	*p*
$(C_6H_5)_2Sn$	Benzene	5 and 6	Vapor pressure osmometry	*q*

[a] J. Berkowitz, D. A. Bafus, and T. L. Brown, *J. Phys. Chem.* **65,** 1380 (1961).
[b] J. J. Brooks and G. D. Stucky, *J. Am. Chem. Soc.* **94,** 7333 (1972).
[c] A. I. Snow and R. E. Rundle, *Acta Cryst.* **4,** 348 (1951).
[d] G. E. Coates and P. D. Roberts, *J. Chem. Soc. A* 2651 (1968).
[e] A. Almenningen, A. Haaland, and J. E. Nilson, *Acta Chem. Scand.* **22,** 972 (1968).
[f] E. C. Ashby and M. B. Smith, *J. Am. Chem. Soc.* **86,** 4363 (1964).
[g] G. D. Stucky and R. E. Rundle, *J. Am. Chem. Soc.* **86,** 5344 (1964).
[h] R. G. Vranka and E. L. Amma, *J. Am. Chem. Soc.* **89,** 3121 (1967).
[i] A. Almenningen, S. Salvorsen, and A. Haaland, *Acta Chem. Scand.* **25,** 1937 (1971).
[j] N. Muller and A. L. Otermat, *Inorg. Chem.* **4,** 296 (1965).
[k] J. P. Oliver and L. G. Stevens, *J. Inorg. Nucl. Chem.* **24,** 953 (1962).
[l] L. M. Dennis, R. W. Work, E. G. Rochow, and E. M. Chamot, *J. Am. Chem. Soc.* **56,** 1047 (1934).
[m] N. Muller and A. L. Otermat. *Inorg. Chem.* **2,** 1075 (1963).
[n] H. Gilman and G. L. Schwebke, *Adv. Organometal. Chem.* **1,** 94 (1964).
[o] C. A. Kraus and W. V. Sessions, *J. Am. Chem. Soc.* **47,** 2361 (1925).
[p] C. A. Kraus and W. V. Sessions, *J. Am. Chem. Soc.* **48,** 2131 (1926).
[q] W. P. Neumann and K. König, *Justus Liebigs Ann. Chem.* **677,** 1 (1964).

4. Hydrolysis and Other Carbon–Metal Bond Cleavages

The hydrolytic cleavage of a carbon–metal bond, followed by the qualitative and quantitative analysis of the resulting hydrocarbon, is central to the identification of new organometallic compounds.[1] In addition, treatment of an organometallic with deuterium oxide and nmr spectroscopic demonstration of the position of the deuteron in the isolated hydrocarbon often locate the position of the carbon–metal bond.[183] This deuterium-labeling procedure has the advantage of being applicable for unisolable solution species of organometallics.[184]

The conditions and apparatus for the hydrolysis of various organometallics depends upon their sensitivity to protic sources. A higher alcohol will suffice to cleave all three bonds of aluminum alkyls at room temperature[185]; gallium alkyls require heating with dilute acid to cleave all alkyl groups[120]; and boron alkyls will be found to cleave completely only in refluxing glacial acetic acid.[186] Too vigorous an initial reaction can be suppressed by strong cooling upon admixing the reactants.[187] When the hydrocarbons are gaseous, they can be collected in a mercury-filled gas buret attached to the apparatus shown in Fig. 10.[185] In the apparatus, the alkyl can be weighed directly into E (stopped at H); or a tared, sealed ampul is placed in E and the setup completed and flushed before the ampul is broken open. If the hydrocarbons formed are liquids, they are mixed with a standard amount of marker liquid and the amount determined by glc. Where more than one hydrocarbon is formed, as in the hydrolysis of an unsymmetric alkyl [R_2BR'],[186] either gas chromatography or mass spectrometry at low accelerating voltages (10 eV) yields the ratio of products.

The lower reactivity of certain organometallics may permit a rather clean cleavage of only certain of the existing carbon–metal bonds in a R_mM unit, and the resulting product may be isolated for purposes of identification. The smooth alcoholysis of gallium alkyls to dialkylgallium alkoxide is illustrative.[120] The isolation of stable cleavage intermediates is the rule rather than exception with organometalloids. A further distinction in

[183] Such labeling technique may give erroneous results with allylic, propargylic, or benzylic organometallics, where protodemetallation may occur with rearrangement; see J. J. Eisch, *Adv. Organometal. Chem.* **16,** 91–98 (1977).

[184] See the detection of thermally labile α,β-epoxyalkyllithium reagents: J. J. Eisch and J. E. Galle, *J. Am. Chem. Soc.* **98,** 4646 (1976).

[185] K. Ziegler, H. G. Gellert, H. Martin, K. Nagel, and J. Schneider, *Justus Liebigs Ann. Chem.* **589,** 91 (1954).

[186] (a) H. C. Brown and K. Murray, *J. Am. Chem. Soc.* **81,** 4108 (1959); (b) L. H. Toporcer, R. E. Dessy, and S. I. E. Green, *J. Am. Chem. Soc.* **87,** 1236 (1965).

[187] If the hydrolysis of alkylaluminum compounds is allowed to occur too rapidly, pyrolysis products may arise. Thus such hydrolysis of triethylaluminum can give rise to ethylene and hydrogen gas.

TABLE IV

EXPERIMENTAL CONDITIONS FOR THE PROTOLYSIS OF CARBON–METAL BONDS

Compound	Reagent	Conditions	Products	Reference
C_6H_5M, M = Li, Na, K, Rb, Cs	H_2O	25°, instantaneous, possibly explosive	C_6H_6	*a, b*
n-C_4H_9Li	THF	25°, gradual	C_4H_{10}, $CH_2{=}CH_2$, $CH_2{=}CHOLi$	*c*
$(CH_3)_2Be$	CH_3OH	−100°, 1 equiv	CH_3BeOCH_3	*d*
CH_3MgI	HA	25°, in $(n\text{-}C_4H_9)_2O$ solution (Zerewitinoff active hydrogen)	CH_4	*e*
$(C_5H_5)_2M$, M = Ca, Sr, Ba	H_2O	25°, rapid	C_5H_6	*f*
$(CH_3)_2Zn$	H_2O	25°, possibly explosive	CH_4, $Zn(OH)_2$	*g*
$(CH_3CH_2)_2Cd$	H_2O	25°, slow	CH_4, $Cd(OH)_2$	*h*
$(CH_3)_3B$	H_2O	25°, N.R.; 180°, slow	CH_4, $(CH_3)_2BOH$	*i*
$(C_6H_5)_3B$	CH_3CH_2OH	25°	C_6H_6, $(C_6H_5)_2BOCH_2CH_3$	*j*
$(CH_3)_3Al$	H_2O	25°, instantaneous, possibly explosive	$CH_4 + Al(OH)_3$	*k*
$(CH_3)_3Ga$	H_2O	0°, 0.5 equiv, rapid;	$[(CH_3)_2Ga]_2O$	*l*
		etherate, 150°, 1 equiv	$(CH_3GaO)_x$	*m*
$(CH_3)_3In$	H_2O	0°, 2 equiv, vigorous	$CH_3In(OH)_2$	*n*
$(CH_3)_3Tl$	H_2O	25°	$(CH_3)_2TlOH$	*o*
R_4Si, R_4Ge	H_2SO_4	0°, concentrated	N.R.	—
$(CH_3)_4Pb$	HCl	25°	$(CH_3)_3PbCl + CH_4$	*p*

[a] W. Schlenk and J. Holtz, *Ber.* **50,** 262 (1917).

[b] H. Gilman and R. V. Young, *J. Org. Chem.* **1,** 315 (1936).

[c] R. B. Bates, L. M. Kroposki, and D. E. Potter, *J. Org. Chem.* **37,** 560 (1972).

[d] G. E. Coates and H. Fishwick, *J. Chem. Soc. A* 477 (1968).

[e] M. S. Kharasch and O. Reinmuth, "Grignard Reactions of Nonmetallic Substances," p. 1166. Prentice-Hall, Englewood Cliffs, New Jersey, 1964.

[f] E. O. Fischer and G. Stölzle, *Chem. Ber.* **94,** 2187 (1961).

[g] E. Frankland and D. F. Duppa, *Justus Liebigs Ann. Chem.* **130,** 118 (1864).

[h] E. Krause, *Ber.* **50,** 1813 (1917).

[i] J. Goubeau and R. Epple, *Chem. Ber.* **90,** 170 (1957).

[j] E. Krause and R. Nitsche, *Ber.* **55,** 1261 (1922).

[k] E. Krause and A. V. Grosse, "Die Chemie der Metallorganischen Chemie," Borntraeger, Berlin, 1937.

[l] G. E. Coates and K. Wade, "Organometallic Compounds," Vol. I, p. 347, Methuen, London, 1967.

[m] M. E. Kenney and A. I. Laubengayer, *J. Am. Chem. Soc.* **76,** 4839 (1954).

[n] L. M. Dennis, R. W. Work, E. G. Rochow, and E. M. Chamot, *J. Am. Chem. Soc.* **56,** 1047 (1934).

[o] H. Gilman and R. G. Jones, *J. Am. Chem. Soc.* **68,** 517 (1946).

[p] R. Heap and B. C. Saunders, *J. Chem. Soc.* 2983 (1949).

organometalloidal hydrolysis is that alkaline media are more effective than water alone. The complete or selective hydrolytic cleavage of a collection of organometallics is compiled for further reference in Table IV. With organometallics undergoing cleavage of their carbon–metal bonds with rearrangement,[183] spectral measurements on the organometallic itself must be used to verify the position of the metal on the carbon skeleton.

Many other reagents, both organic and inorganic, can be used to verify the presence of carbon–metal bonds but not always their actual site or stereochemistry. An extensive discussion of the scope and limitations of these reactions is available.[1] For convenience to the experimenter, such derivatization procedures leading to high yields are summarized with key references in Table V. Those capable of quantitative application in organometallic analysis or reaction kinetics are marked with an asterisk. Specific attention should be given to the methods for the titrimetric analysis of Grignard and organolithium reagents, since they are frequently used operations in organometallic synthesis. For recent reviews of these analytical methods, see refs. 132c and 175.

TABLE V

REAGENTS FOR THE CLEAVAGE AND DERIVATIZATION OF CARBON–METAL BONDS

Compound	Reagent	Applicability	Product	Reference
R_nM	D_2O or $R'OD$	M = Groups IA, IIA, Zn, Al	R—D*	*a*
R_nM	CO_2	M = Group IA, Mg	$R—CO_2H$*	*b*
RLi or RMgX	$(CH_3)_3SiCl$	Quantitative	$R—Si(CH_3)_3$*	*c*
RLi	$(C_6H_5)_2CO$	Facile isolation	$R—C(C_6H_5)_2OH$*	*d*
R_nM	I_2	M = Groups IA, IIA, IIB, Al, Ga, In, Tl	RI (complex reaction)	*e*
R_3B	H_2O_2–NaOH	Most boranes	ROH	*f*
R_3B	$(CH_3)_3NO$	Quantitative	$B(OR)_3$*	*g*
R_3Al	O_2	Not satisfactory for Ar_3Al	$Al(OR)_3$	*h*

[a] G. Wilke and H. Müller, *Justus Liebigs Ann. Chem.* **618,** 267 (1958).

[b] H. Ruschig, R. Fugmann, and W. Meixner, *Angew. Chem.* **70,** 71 (1958).

[c] H. Gilman, R. A. Benkeser, and G. E. Dunn, *J. Am. Chem. Soc.* **72,** 1689 (1950); H. O. House and W. L. Respess, *J. Organometal. Chem.* **4,** 95 (1965).

[d] G. Wittig, R. Ludwig, and R. Polster, *Chem. Ber.* **88,** 294 (1955).

[e] A. F. Clifford and R. R. Olsen, *Anal. Chem.* **32,** 544 (1960).

[f] G. Zweifel and H. C. Brown, *in* "Organic Reactions" (A. C. Cope, ed.), Vol. XIII, p. 1, Wiley, New York, 1963.

[g] R. Köster and Y. Morita, *Justus Liebigs Ann. Chem.* **704,** 70 (1967).

[h] K. Ziegler, F. Krupp, and K. Zosel, *Justus Liebigs Ann. Chem.* **629,** 241 (1960); G. A. Razuvaev, E. V. Mitrofanova, and G. G. Petukhov, *Zh. Obsh. Chim.* **30,** 1996 (1960).

5. Spectroscopy

The previous observations on practical spectroscopy given in Volume I need only be supplemented here by a brief discussion of suitable measuring cells for the highly air- and moisture-sensitive organometallics of main-group metals. Furthermore, structural features arising from the presence of the metal are reflected in certain characteristic spectroscopic properties. The utility of these special spectroscopic features will be noted as an aid to characterization of the organometallic.

a. Methods

The loading of a sample tube for nmr spectroscopic examination can be readily performed by placing the sample tube in the apparatus shown in Fig. 33, flushing, and filling with compound and solvent. Airtight Teflon plugs can be used, or the plastic cap can be sealed with tape. The rather high concentration used does not cause great concern about small amounts of contamination. But the major difficulty is finding a solvent that neither destroys nor solvates the organometallic. Some suitable solvents for specific organometallics are compiled in Table I.

Samples for infrared study may be prepared by filling demountable, screw-thread cells in a well-functioning inert atmosphere chamber. Such a course must be followed if mulls are to be employed. In this case, the agate mortar and pestle and thoroughly dried (heated with molten sodium), deaerated mineral oil are stored and used in the chamber. For liquids a setup that has solved the problem of measuring the reactive aluminum alkyls and hydrides has been reported.[188] A modified desiccator can be used for preparing infrared spectral samples (Fig. 33). After the sample is introduced by syringe, long forceps can be used to cover it before exposure to the atmosphere.

The reliable measurement of ultraviolet spectra of reactive organometallics is very demanding because the low concentrations make the disturbing effect of impurities from hydrolysis, oxidation, or surface elution loom large. An adaptation of the foregoing setup, permitting continual renewal of the cell contents, is a useful expedient.[189] Because of association equilibria and the Lewis acid character of many metal alkyls, severe deviations from Beer's law may rule out quantitative conclusions. In less sensitive systems, ordinary

---

[188] E. G. Hofmann and G. Schomburg, *Z. Elektr. Chem.* **61,** 1101 (1957) and Houben-Weyl: "Methoden der Organischen Chemie," Band XIII/4, p. 300. Georg Thieme Verlag, Stuttgart, Germany, 1970.

[189] W. P. Neumann, *Justus Liebigs Ann. Chem.* **629,** 23 (1960).

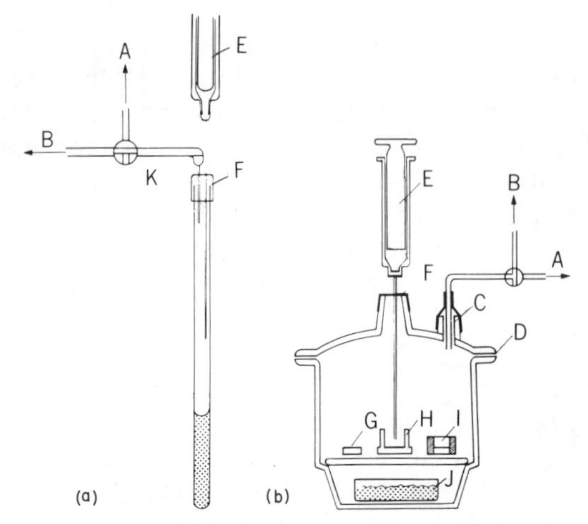

FIG. 33. Procedures for preparing liquid organometallic samples for spectroscopic measurements. (a) Nmr sample. A, Vacuum; B, inert gas; E, gastight syringe; F, septum; K, purging needles. (b) Infrared sample. C, Snug but flexible rubber sleeve; D, tight-fitting flange-top vacuum desiccator; G, salt optics; H, infrared cell support; I, infrared sleeve-type cover; J, P_2O_5 desiccant.

ultraviolet cuvets may be equipped with rubber stoppers and purged or filled with syringe needles inserted through the rubber seal (Fig. 33).[190]

b. CHARACTERISTIC ABSORPTIONS

Compiled in Tables VI–IX are the spectroscopic absorptions peculiar to organometallics of the main group, hence most useful in their identification. The lack of specific character for ultraviolet absorptions in general and the possible appearance of charge transfer bands in certain solvents make these bands of decidedly less value in identification. In contrast, the wealth of specific qualitative and quantitative detail in a given nmr spectrum makes it the most valuable. However, because of the tendency of Group IA–IIIA alkyls to act as Lewis acids toward themselves or toward electron-pair donors, pronoünced solvent, concentration, and temperature effects are to be expected in such spectra.

---

[190] See R. Waack and M. A. Doran, *J. Am. Chem. Soc.* **85,** 1651 (1963), for a quartz absorption cell apparatus sealed to the side arm of a two-necked, round-bottomed flask and suitable for measuring electronic spectra of organolithium compounds.

TABLE VI

PROTON MAGNETIC RESONANCE SPECTRAL DATA FOR MAIN-GROUP ORGANOMETALLIC COMPOUNDS[a,b]

Compound	Proton grouping	Chemical shift, δ (ppm)	Coupling constant, J (Hz)
CH_3Li (in ether)	CH_3	-1.3	—
CH_3MgI or $(CH_3)_2Mg$ (in ether)	CH_3	-1.3	—
$(CH_3)_2Zn$ (in C_6H_6)	CH_3	0.53	—
$(CH_3)_2Cd$ (in C_6H_6)	CH_3	1.12	—
$(CH_3)_2Hg$ (in C_6H_6)	CH_3	2.0	—
$(CH_3)_3B$	CH_3	0.8	—
$[(CH_3)_3Al]_2$ (at $-75°$)	CH_3 (bridging)	0.3	—
	CH_3 (terminal)	-0.8	—
$(CH_3)_4Ge$	CH_3	0.12	—
$(CH_3)_4Sn$	CH_3	0.07	—
$(CH_3)_4Pb$	CH_3	0.70	—
CH_3CH_2Li (in C_6H_6)	CH_2	-0.8	8.4
	CH_3	1.88	8.4
$(CH_3CH_2)_4Si$	CH_2	0.6	7.9
	CH_3	1.0	7.9
$CH_2{=}CHMgCl$ (in THF)	CH_α	6.68	17.6 (α, c)
	CH_{β_t}	6.13	23.0 (α, t)
	CH_{β_c}	5.57	7.5 (c, t)
$C_6H_5C(H){=}C(H)Si(CH_3)_3$[c] (both H above)	CHSi	5.83	15
	CHC_6H_5	7.36	15
$C_6H_5C(H){=}C(H)Si(CH_3)_3$[c] (H above and below)	CHSi	6.33	19
	CHC_6H_5	6.88	19
$CH_2{=}CHCH_2MgBr$	CH_2	4.8	12
	CH	7.3	12
$(C_6H_5)_4M$, M = Si, Sn	CH (o)	7.6	—
	CH (m, p)	7.4	—
$[C_6H_5C{\equiv}CAl(C_6H_5)_2]_2$[d]	CH (o)	8.37	—
	CH (m, p)	7.55	—
$C_6H_5C{\equiv}CSn(CH_3)_3$[c]	CH (o)	7.15	—
	CH (m, p)	7.05	—

[a] See the compilation of nmr spectral data by M. L. Maddox, S. L. Stafford, and H. D. Kaesz, *Adv. Organometal. Chem.* **3,** 1–179 (1965).

[b] See the updating of nmr spectral data by R. G. Kidd, *in* "Characterization of Organometallic Compounds" (M. Tsutsui, ed.), Part II, pp. 373–480, Wiley (Interscience), New York, 1971.

[c] J. J. Eisch and M. W. Foxton, *J. Org. Chem.* **36,** 3520 (1971).

[d] J. J. Eisch and W. C. Kaska, *J. Organometal. Chem.* **2,** 184 (1964).

TABLE VII

NUCLEAR MAGNETIC RESONANCE SPECTRAL DATA FOR SIGNIFICANT HETEROATOMS OTHER THAN HYDROGEN IN MAIN-GROUP ORGANOMETALLICS

Nucleus	Isotope	Natural abundance (%)	Spin quantum number	Reference
Li	7	92.57	$\frac{3}{2}$	*a*
B	10	18.83	3	*b*
B	11	81.17	$\frac{3}{2}$	*b*
C	13	1.11	$\frac{1}{2}$	*c*
Al	27	100	$\frac{5}{2}$	*a*
Si	29	4.7	$\frac{1}{2}$	*d*
Cd	111	12.86	$\frac{1}{2}$	*e*
Cd	113	12.34	$\frac{1}{2}$	*e*
Sn	117	7.67	$\frac{1}{2}$	*f*
Sn	119	8.68	$\frac{1}{2}$	*f*
Hg	199	16.86	$\frac{1}{2}$	*a*
Tl	203	29.52	$\frac{1}{2}$	*g*
Tl	205	70.48	$\frac{1}{2}$	*g*
Pb	207	21.11	$\frac{1}{2}$	*h*

[a] R. G. Kidd, *in* "Characterization of Organometallic Compounds" (M. Tsutsui, ed.), Vol. II, p. 373, Wiley (Interscience), New York, 1971.

[b] G. R. Eaton and W. N. Lipscomb, "NMR Studies of Boron Hydrides and Related Compounds," Benjamin, New York, 1969.

[c] J. B. Stothers, "Carbon-13 NMR Spectroscopy," Academic Press, New York, 1972; E. Breitmaier and W. Voelter, "Carbon-13 NMR Spectroscopy," 2nd ed., Verlag Chemie, Weinheim, Germany, 1978.

[d] G. Englehardt, H. Jancke, M. Mägi, T. Pehk, and E. Leppma, *J. Organometal. Chem.* **28,** 293 (1971).

[e] A. D. Cardin, P. D. Ellis, J. D. Odom, and J. W. Howard, Jr., *J. Am. Chem. Soc.* **97,** 1672 (1975).

[f] P. G. Harrison, S. E. Ulrich, and J. J. Zuckerman, *J. Am. Chem. Soc.* **93,** 5398 (1971).

[g] J. P. Maher and D. F. Evans, *J. Chem. Soc.* 637 (1965).

[h] G. E. Maciel and J. L. Dallas, *J. Am. Chem. Soc.* **95,** 3039 (1973).

TABLE VIII
CHARACTERISTIC INFRARED SPECTRAL ABSORPTION FREQUENCIES FOR MAIN-GROUP ORGANOMETALLICS

Compound	Frequency (cm^{-1})	Grouping	Reference
$[CH_3Li]_4$	300–600	C—Li	*a*
$(CH_3)_3B$	1150	C—B	*b*
$[(CH_3)_3Al]_2$	479, 565, 696	C—Al	*c*
$(CH_3)_4Si$	696	C—Si	*d*
$(CH_3)_4Ge$	602	C—Ge	*e*
$(CH_3)_4Sn$	528	C—Sn	*e*
$(CH_3)_4Pb$	478	C—Pb	*e*
$[R_2BH]_2$	1565	B—H	*f*
$[R_2AlH]_3$	1750 (br)	Al—H	*g*
$(C_6H_5)_3SiH$	2095	Si—H	*h*
$(C_6H_5)_3SnH$	1850	Sn—H	*i*
$(CH_3)_3SiC{\equiv}CC_6H_5$	2180	C≡C	*j*
$(CH_3)_3SnC{\equiv}CC_6H_5$	2140	C≡C	*j*
$(C_6H_5)_3MCH{=}CH_2$, M = Si, Ge, Sn, Pb	942–960	=C—H	*k*

[a] R. West and W. Glaze, *J. Am. Chem. Soc.* **83,** 3580 (1961).

[b] W. J. Lehman, C. O. Wilson, and I. Shapiro, *J. Chem. Phys.* **28,** 781 (1958).

[c] T. Ogawa, K. Hirota, and T. Miyazawa, *Bull. Chem. Soc. Jpn.* **38,** 1105 (1965); E. G. Hoffmann, *Z. Elektrochem.* **64,** 616 (1960).

[d] I. Simon and H. O. McMahon, *J. Chem. Phys.* **20,** 905 (1952).

[e] E. R. Lippincott and M. C. Tobin, *J. Am. Chem. Soc.* **75,** 4141 (1953).

[f] R. Köster, G. Bruno, and P. Binger, *Justus Liebigs Ann. Chem.* **644,** 1 (1961).

[g] G. Schomburg and E. G. Hoffmann, *Z. Elektrochem.* **61,** 1110 (1957).

[h] "Sadtler Standard Spectra," Infrared Spectrum No. 12209, Sadtler Research Laboratories, Philadelphia, Pennsylvania.

[i] W. P. Neumann, *Angew Chem.* **75,** 225 (1963).

[j] J. J. Eisch and M. W. Foxton, *J. Org. Chem.* **36,** 3520 (1971).

[k] K. Nakamoto, *in* "Characterization of Organometallic Compounds" (M. Tsutsui, ed.), pp. 77–80. Part I, Wiley (Interscience), New York, 1969.

TABLE IX

CHARACTERISTIC ULTRAVIOLET AND VISIBLE SPECTRAL DATA FOR MAIN-GROUP ORGANOMETALLICS

Compound and solvent	Band (nm) and intensity (log ε)	Reference
n-Butyllithium (hexameric in C_6H_6)	180, 221, 225 (br)	*a*
Phenyllithium (THF)	261 (3.02), 268 (2.99), 292 (2.88)	*b*
Benzyllithium (THF)	330 (3.98)	*c*
9-Fluorenyllithium (dioxane)	346, 437 (contact ion pairs)	*d*
9-Fluorenyllithium (cyclohexylamine)	452 (3.03), 477 (3.11), 510 (2.92) (solvent-separated ion pairs)	*e*
Phenylmagnesium bromide (THF)	255	*f*
Trivinylborane (gas)	234 (4.3)	*g*
Triphenylborane (methylcyclohexane)	238 (4.28), 276 (4.54), 287 (4.59)	*h*
Tetraphenylsilane ($CHCl_3$)	254 (3.04), 260 (3.11), 265 (3.15), 272 (3.04)	*i*
Tetraphenylgermane (heptane)	252 (3.32), 259 (3.23), 263 (3.09), 269 (2.93)	*j*
Tetraphenylstannane (heptane)	247 (2.09), 252 (3.06), 259 (3.11), 262 (2.96), 265 (3.08), 268 (2.74)	*j*
Tetraphenylplumbane (ethanol)	250, 257, 263, 268 (sh)	*k*
Triphenylarsine (ethanol)	248 (4.12), 273 (sh), 277 (sh)	*l*
Triphenylstilbine (ethanol)	256 (4.07)	*l*
Triphenylbismuthine (ethanol)	248 (4.11), 280 (sh, 3.63)	*l*
Pentaphenylborole (toluene)	567	*m*
Heptaphenylborepin (ether)	249, 280, 345, 425	*m*

[a] J. P. Oliver, J. B. Smart, and M. T. Emerson, *J. Am. Chem. Soc.* **88,** 4101 (1966).
[b] R. Waack and M. A. Doran, *J. Am. Chem. Soc.* **85,** 1651 (1963).
[c] R. Waack and M. A. Doran, *J. Phys. Chem.* **67,** 148 (1963).
[d] T. E. Hogen-Esch and J. Smid, *J. Am. Chem. Soc.* **88,** 307 (1966).
[e] A. Steitwieser, Jr., and J. I. Brauman, *J. Am. Chem. Soc.* **85,** 2633 (1963).
[f] G. Fraenkel, S. Dayagi, and S. Kobayashi, *J. Phys. Chem.* **72,** 953 (1968).
[g] C. D. Good and D. M. Ritter, *J. Am. Chem. Soc.* **84,** 1162 (1962).
[h] B. G. Ramsey, *J. Phys. Chem.* **70,** 611 (1966).
[i] G. Milazzo, *Gazz. Chim. Ital.* **71,** 73 (1941).
[j] S. R. LaPaglia, *J. Mol. Spectrosc.* **7,** 427 (1961).
[k] C. N. R. Rao, J. Ramachandran, and A. Balasubramanian, *Can. J. Chem.* **39,** 171 (1961).
[l] H. H. Jaffe, *J. Chem. Phys.* **22,** 1430 (1954).
[m] J. J. Eisch and J. E. Galle, *J. Am. Chem. Soc.* **97,** 4436 (1975).

E. Safety Measures

1. General Precautions

Although the safe manipulation of organometallics was stressed throughout the previous sections, certain general hazards of research in this area need emphasis.[191] When one considers how the dangers of organic chemistry itself are accentuated by compounds that are at once explosively hydrolyzable and oxidizable, and often toxic, there is no other *modus vivendi* than circumspect technique. Maintain the pessimistic view that the worst could happen and then provide for this eventuality. The following suggestions are worth the extra effort:

a. Do not try to save time by manipulating large quantities of pure, self-igniting organometallics, or in large volumes of flammable solvents, at once.

b. Wear safety glasses and a face shield, as well as asbestos leather-faced gloves with elbow-length gauntlets, when transferring ignitable alkyls.

c. Place all containers, storage tanks, and receivers, during transfers, in a metal trough able to contain the liquid in case of breakage or spillage.

d. Keep all solvents and water bottles out of the area of alkyl work.

e. Have clean, dry sand, vermiculite, or diatomaceous earth ready to cast on a fire; a solid sodium cabonate nozzle fire extinguisher is handy to direct onto a blaze.

f. Large quantities of metal alkyls may be disposed of in a dump, but humanity demands that such compounds in sealed bottles be ruptured from a safe distance and not left closed to be discovered by some unknowing victim.

g. Smaller quantities of metal alkyls should be disposed of by diluting first with xylene and then by the slow addition of a higher alcohol (e.g., amyl alcohol), care being taken to provide for gas evolution in certain cases.

h. All tubes containing reactive alkyls (e.g., pipets) should be washed out with xylene immediately after use, lest a packet of liquid alkyl be sealed off by oxidation crusts and then released violently when the tube is washed.

---

[191] Additional guidance on the hazards and safe manipulation of these compounds can be obtained from the following sources: (*a*) product bulletins from the Ethyl Corporation, the Lithium Corporation of America, MSA Research Corporation, and Texas Alkyls; and (*b*) introductory discussion of manipulation in the Houben-Weyl volumes, "Methoden der Organischen Chemie," which cover organometallic compounds.

i. Hydrolysis of any reactive mixture or distillation residue containing metal alkyls, metal hydrides, or active metals should be done right after use and with caution (see point g). Many solvents, even in the presence of hydrides or metals, slowly form dangerous peroxides. If hydrogen is evolved, a sparkless stirring motor is essential.

j. Certain metal alkyls or their decomposition products are toxic if inhaled or absorbed through the skin, especially those of beryllium, boron, mercury, thallium, and lead. Special hoods having a strong, independent exhaust system should be reserved for such manipulations. The operator should take pains not to spill or spread residues around outside the hood and should always wear impervious gloves.

Although vapors must of necessity be exhausted into such hoods, any toxic liquid alkyl should be chemically destroyed and the resulting inorganic salt disposed of in an environmentally acceptable fashion (Section E,2).

2. Toxicity Dangers

The recent public awareness of the adverse effect of chemicals on the environment has had a beneficial influence on acceptable laboratory practice.[192] Long accustomed to working with substances of proven or probable toxicity, laboratory chemists had generally become cavalier, if not careless, about possible long-term exposure. With the sobering realization of the pernicious, cumulative effect of many common laboratory reagents, chemists now employ better ventilation, disposal, and skin protection techniques than formerly.

In working with organometallics known to be highly toxic, such as metal carbonyls and the volatile alkyls of beryllium, mercury, thallium, and lead, one needs little encouragement to take elaborate precautions and possibly even to set up an isolated, properly vented hood working area. For common organometallics, quantitative information on any known toxic effects may be found in the most recent volume of "The Toxic Substance List," compiled by the National Institute for Occupational Safety and Health.[193] But the toxic effects, especially those of a long-term or cumulative kind, of most organometallics are either unknown or have not been investigated in a quantitative manner. Moreover, since most organometallics undergo hydrolysis, oxidation, or complexation (Section A,1), the possible toxicity of inter-

---

[192] See J. M. Wood's article, Biological cycles for toxic elements in the environment, *Science* **183**, 1049 (1974).

[193] H. E. Christensen and T. T. Luginbyhl (eds.), "The Toxic Substances List," p. 904. U.S. Dept. Health, Education and Welfare, Rockville, Maryland, 1974. An ongoing registry of toxic effects of chemical substances is maintained by the National Institute for Occupational Safety and Health (NIOSH).

mediates arising from these processes must also be considered.[194] Because of all these uncertainties, prudent practice dictates that all organometallic vapors be absorbed into solution or condensed in cold traps, that scrupulous neatness be exercised toward spillage, that such organometallic waste, as well as solid residues, be treated chemically to yield inorganic salts, and that such salts be disposed of properly.[195] If costs warrant the effort (gallium, silver, or gold), naturally chemical recovery is in order.

3. Fire Hazards

The exothermic air-oxidation of the volatile alkyls of Groups I–III makes them extremely pyrophoric. Although pure higher alkyl homologs may not ignite in air, concentrated solutions or suspensions in volatile hydrocarbons or ethers may inflame in air. *Especially the addition of metal alkyls or hydrides to peroxide-containing solvents can lead to a spontaneous fire or explosion* (Section A,3).

Highly exothermic organometallic reactions, such as protolysis, oxidation, or even addition reactions (ethylene insertion in a carbon–aluminum bond or donor complexation), should be carried out with proper control of admixing and removal of the heat of reaction, lest spontaneous ignition of the solvent or a violent reaction ensue. Although such a reaction is conducted under an inert atmosphere, a runaway reaction may spew the reaction contents into the air. To avoid such an outcome, cooling baths and condensers equipped with an inert coolant should be used. Also, the reactants should be suitably diluted with an inert solvent to moderate the vigor of the reaction and to provide a convenient reflux temperature for heat removal. In addition, the whole reaction vessel should be contained in a metal tray deep enough to hold the reaction mixture in case of breakage.

4. Explosions

As mentioned in the foregoing section, any highly exothermic reaction involving volatile components can lead to an explosion. The sudden release of gas, especially hydrogen, can be particularly hazardous. The accidental protolysis or thermolysis of organometallic hydrides (Section C,1) should be scrupulously guarded against. Fragile water lines (rubber or glass tubing) should be excluded from organometallic storage or reaction areas. Heating

---

[194] Under contract from NIOSH this author has made a survey of organometallic compounds used either in industry and commerce or in development and research laboratories. The toxicity of these approximately 1100 organometallics and their degradation products is undetermined.

[195] See "Laboratory Waste Disposal Manual," pp. 176. Manufacturing Chemists Association, 1825 Connecticut Ave., N.W., Washington, D.C., 20009, 1975.

devices for reaction or distillation procedures should never be left unattended.

Contact between organometallics and organic peroxides presents a most serious hazard (Section A,3). There are occasions where, of necessity, such contact is encountered, as in the air-oxidation of metal alkyls for the purpose of preparing alcohols[187] or hydroperoxides.[188] In such instances, extreme caution is mandatory, especially if metal peroxide intermediates accumulate. The formation of solid metal peroxide crusts, due to solvent evaporation by the airstream, can lead to explosion, for such solids are shock-sensitive.

PART II

Specific Preparations of Nontransition-Metal Organometallic Compounds

A. Introduction

1. Criteria Governing the Choice of Procedures

This part presents detailed procedures for the preparation of about 85 nontransition-metal organometallic compounds. As a working definition, we view a transition metal as an element having partly filled *d* or *f* shells in any commonly occurring oxidation state.[1] Accordingly, compounds of Group IIB (zinc, cadmium, and mercury) are included; compounds of Group IB (copper, silver, and gold) are not. The concept of a nontransition or main-group metal is broadened to include not only Groups IA, IIA, and IIIA, but also the representatives of Groups IVA, VA, VIA, and VIIA known as metalloids.[2] Subsequent volumes in this series will offer preparations of such compounds as those of selenium, tellurium, and iodine. Although there are distinct differences in chemical behavior between main-group organometallics and these organometalloids,[3] their common feature is a polar, often weak, carbon–element bond. Hence the formation and the cleavage reactions of this bond constitute the essential chemistry of both organometallic and organometalloid compounds.

The methods offered here illustrate either the preparation of generally useful structural types or the application of important synthetic transformations. In choosing procedures, we have generally avoided those for commercially available and relatively inexpensive organometallic reagents and those already described in "Inorganic Syntheses" or "Organic Syntheses." Such preparations are included here only where modifications in procedure have led to a clear advantage in ease, yield, or purity. Where several methods are available, we have been swayed in favor of those with which our group has had successful experience. Generally, these methods have been tested by two or more chemists.

Because the chemical behavior of main-group organometallics is principally governed by the nature of the metal,[3] the following preparations are grouped by periodic family and then by specific metal. This classification

[1] F. A. Cotton and G. Wilkinson, "Advanced Inorganic Chemistry," 3rd ed., p. 528. Wiley (Interscience), New York, 1972.

[2] E. G. Rochow, "The Metalloids," Heath, Boston, Massachusetts, 1966.

[3] J. J. Eisch, "The Chemistry of Organometallic Compounds," p. 3. Macmillan, New York, 1967.

differs from that of Volume 1, where the nature of the ligands determines the groupings. Nevertheless, we have tried here to illustrate the due importance of different carbon ligands by choosing various R_mM derivatives, such as alkyl, vinylic, allylic, acetylenic, aryl, benzylic, functionally substituted R, metallocyclic, and dimetallo types. Also, examples of alkoxy, halo, hydrido, and solvate (ether, amine) ligands are given.

2. General Procedural Observations

The kind of standard apparatus required for each preparation will be indicated by a reference to Part I. Where specialized apparatus has been suggested for alternative procedures, the reader is referred to the original literature. Unless otherwise stated, all reactions should be conducted under an atmosphere of anhydrous, oxygen-free nitrogen. Conventions in presenting data will be those currently used by journals of the American Chemical Society. For spectral data, nmr proton chemical shifts are given in parts per million on the δ scale with reference to internal tetramethylsilane, and coupling constants are in hertz units; all temperatures are in degrees Celsius; infrared absorptions are in reciprocal centimeters; and ultraviolet peaks in nanometers.

Hazards are inherent in most organometallic operations, and a careful reading of Part I by the inexperienced should precede any laboratory work. A special caveat should be heeded for those reactions involving heterogeneous reactants (e.g., a metal and an organic halide). *Be certain that the reaction has started before large amounts of the reagents are admixed.* Otherwise, a long induction period may produce conditions for an uncontrolled or even explosive reaction. After introducing small amounts of one reagent to the other, look for signs of reaction (color, heat, turbidity) to gain assurance of proper initiation.

B. Compounds of Group IA

1. General Preparative and Analytical Methods

Although the organometallic derivatives of cesium, rubidium, potassium, and sodium all surpass organolithium compounds in reactivity, their lower solubility in alkanes and their rapid attack on almost any other organic dispersing medium make them unsuitable for many organometallic reactions. In contrast, many organolithium compounds are either soluble in alkanes or have reasonable kinetic stability in ether or amine solution. Consequently, they can be prepared in solution and employed for a wide variety of organic or organometallic transformations.[4,5] With the exception of other alkali halides, almost any metallic halide, alkoxide, or similar salt can be converted into an organometallic derivative by the appropriate organolithium reagent[6]:

$$RLi + MZ_n \longrightarrow RMZ_{n-1} + LiZ \quad (1)$$

Furthermore, organolithium derivatives add to a most extensive array of unsaturated organic functional groups, ranging from C≡C and C=C to C=N and C=O linkages.

The generally applicable methods of preparing organolithium compounds involve the interaction of lithium metal or an organolithium reagent with hydrocarbons, organic halides, ethers, or organometallics of less active metals. Where practical, the organolithium starting material is chosen from those commercially available, such as methyl-, *n*-butyl-, *sec*-butyl-, or *tert*-butyllithium,[7] or those easily prepared in the laboratory, such as methyl- or phenyllithium in ethyl ether.[8] The following illustrative equations serve

---

[4] U. Schöllkopf, *in* "Methoden der Organischen Chemie" (E. Müller ed.), Vol. XIII/1, Organometallic Compounds, pp. 3–26, 87–254. Georg Thieme Verlag, Stuttgart, Germany, 1970.

[5] B. J. Wakefield, "The Chemistry of Organolithium Compounds," Pergamon, Oxford, 1974.

[6] Especially with transition metals, the RMZ_{n-1} intermediate may have only a transitory existence because of homolytic cleavage of the carbon–metal bond; see F. A. Cotton, *Chem. Rev.* **55,** 551 (1955) for a review of the many early attempts to prepare such labile aryl and alkyl derivatives of transition metals.

[7] See Appendix II for a list of commercial suppliers of organolithium reagents.

[8] These organolithium reagents do not cleave ethyl ether readily, hence can be prepared in a manner analogous to Grignard procedures: see H. Gilman and J. M. Morton, Jr., *in* "Organic Reactions" (R. Adams, ed.), Vol. VIII, p. 286. Wiley, New York, 1954.

to indicate the scope of these methods, and the detailed procedures will show some typical applications.

Lithium metal:

$$RX(Cl, Br, I, OR') + 2Li \longrightarrow RLi + LiX\downarrow \text{ (in alkane)} \quad (2)$$

$$R_nM(Hg, Sn, Pb) + nLi \longrightarrow nRLi + M\downarrow \quad (3)$$

$$R{-}R \text{ (weak C—C)} + 2Li \longrightarrow 2RLi \quad (4)$$

$$2RC{\equiv}CR + 2Li \longrightarrow Li{-}\overset{R}{C}{\cdots}\overset{R}{C}{-}\overset{R}{C}{\cdots}\overset{R}{C}{-}Li \quad (5)$$

$$2RH + 2Li + R'H \text{ (unsat.)} \longrightarrow 2RLi + R'H_3 \quad (6)$$

Alkyllithium (R = aryl, vinyl, alkynyl):

$$RH + R'Li \longrightarrow RLi + R'H \quad (7)$$

$$RX \text{ (X = Br, I)} + R'Li \longrightarrow RLi + R'X \quad (8)$$

$$R_nM \text{ (Hg, Sn, Pb)} + nR'Li \longrightarrow nRLi + R'_nM \quad (9)$$

$$RC{\equiv}CR + R'Li \longrightarrow R{-}\overset{R'}{C}{\cdots}\overset{Li}{C}{-}R \quad (10)$$

Quantitative determination of the concentration of various organolithium reagents in solution can often be achieved by simple hydrolysis of an aliquot and titration of the resulting alkali with 0.1 *N* hydrochloric acid in the presence of a phenolphthalein, methyl orange, or methyl red indicator:

$$RLi + H_2O \longrightarrow RH + LiOH \quad (11)$$

$$ROLi + H_2O \longrightarrow ROH + LiOH \quad (12)$$

Such an analysis is of sufficient accuracy for freshly prepared ethereal solutions of less reactive organolithium compounds, such as methyl-, phenyl-, and 1-alkynyllithium, where contamination with lithium alkoxide [originating from air-oxidation or ether cleavage, Eq. (12)] is minor. Even with solutions of reactive lithium alkyls in alkanes, simple titration of a clear aliquot may suffice, because any adventitiously formed ROLi will precipitate from solution.[4]

Where the simultaneous presence of RLi and ROLi in a sample must be considered, some modification of the Gilman double-titration method must be employed.[9] In essence, a reactive halide, such as benzyl chloride,[9] allyl bromide,[10] or ethylene bromide,[11] is introduced first to react selectively with the reactive organolithium component. Then the resulting mixture is hydrolyzed, and the alkali arising from the remaining ROLi is titrated with acid (*B* milliliters: *y*). This value of required acid is subtracted from that for

[9] H. Gilman and A. H. Haubein, *J. Am. Chem. Soc.* **66,** 1515 (1944).

[10] (a) D. E. Applequist and D. F. O'Brien, *J. Am. Chem. Soc.* **85,** 743 (1963); (b) H. Gilman and F. K. Cartledge, *J. Organometal. Chem.* **2,** 447 (1964).

[11] G. Wittig and G. Harborth, *Chem. Ber.* **77,** 320 (1944).

simply hydrolyzing an identical lithium aliquot and titrating with acid (A milliliters: $x + y$). The value of $A - B$ represents the acid required for the organolithium reagent:

$$(x + y)LiOH \xleftarrow{H_2O} xRLi + yROLi \xrightarrow[-R-R']{R'X} yROLi \xrightarrow{H_2O} yLiOH \qquad (13)$$

Thus, with a nitrogen-flushed pipet or syringe, a 1 to 2-ml aliquot of the organolithium solution is withdrawn and added to 10 ml of distilled water. After thorough shaking the hydrolysate is titrated with 0.1 N hydrochloric acid using a phenolphthalein indicator (A milliliters). Then an identical organolithium aliquot is slowly added to a 125-ml Erlenmeyer suction flask whose side arm is attached to a nitrogen stream and which has been previously flushed with nitrogen and charged with 1 ml of pure allyl bromide or ethylene bromide in 10 ml of anhydrous ethyl ether. After 2–5 min the distilled water (10 ml) is added, and the biphasic mixture thoroughly shaken while titrating as before with the standard acid (B milliliters). For a critique of various analytical procedures, the original papers[9–11] and the excellent review by Schöllkopf[4] should be consulted. In special instances the chemical derivatizations discussed in Part I, Section D,4 (CO_2, Me_3SiCl, Ph_2CO, etc.) can be used for quantitative analysis.

2. Allyllithium

The generally heightened reactivity of allyl- over alkyl- or arylorganometallics in addition reactions makes allyllithium and allyl Grignard reagents most useful in synthesis. The preparation of either allyl derivative directly from allyl halide and metal encounters strong competition from the Wurtz coupling yielding biallyl. Although with magnesium experimental design can surmount this drawback (Section F,19), this approach has not been as useful with lithium. To obviate the chance for Wurtz coupling, two methods have proved feasible for obtaining allyllithium: (*a*) the cleavage of allyl phenyl ether by lithium in THF[12] and (*b*) the tin–lithium exchange between allyltins and phenyllithium.[13] The first method is convenient and appropriate for organic synthesis; the second is better for many organometallic syntheses and for situations requiring the isolation of pure allyllithium.

Procedure A:

$$CH_2{=}CHCH_2{-}O{-}C_6H_5 + 2Li \xrightarrow{THF} CH_2{=}CHCH_2Li + Li{-}O{-}C_6H_5 \qquad (14)$$

In a 500-ml three-necked flask equipped as in Fig. 19 are placed 50 ml of anhydrous THF and 4.2 g (600 mg-atoms) of finely and freshly cut lithium

[12] J. J. Eisch and A. M. Jacobs, *J. Org. Chem.* **28,** 2145 (1963).

[13] D. Seyferth and M. A. Weiner, *J. Org. Chem.* **26,** 4797 (1961).

wire. After the stirred suspension is cooled in a bath at $-15°$, a solution of 6.7 g (50 mmoles) of allyl phenyl ether in 25 ml of anhydrous ethyl ether is added dropwise over 45 min.[14] The cooling bath is removed, and the dark-red suspension stirred for another 15 min. The solution is drained or decanted into a nitrogen-flushed addition funnel. Analysis by the double-titration method gives yields of 62–66%. Assays of yields by derivative formation with benzophenone anil or chloro(triphenyl)silane give isolated yields of pure products in the range of 66–68%.[12]

Procedure B:

$$CH_2{=}CHCH_2{-}Sn(C_6H_5)_3 + C_6H_5Li \longrightarrow CH_2{=}CHCH_2Li + (C_6H_5)_4Sn\downarrow \qquad (15)$$
$$CH_2{=}CHCH_2Li + (n\text{-}C_4H_9)_3SnCl \longrightarrow (n\text{-}C_4H_9)_3SnCH_2CH{=}CH_2 + LiCl$$

In a 500-ml three-necked flask equipped as in Fig. 19 (without drain) are placed 30 g (77 mmoles) of allyl(triphenyl)tin (Section F,19) and 150 ml of anhydrous ethyl ether. The stirred solution is then treated dropwise with 75 ml of a 1.15 *M* solution of phenyllithium in ether (86 mmoles) to cause immediate precipitation of tetraphenyltin. After stirring for 30 min more, a solution of 25 g (77 mmoles) of chloro(tributyl)tin in 50 ml of ether is added dropwise, and the resulting suspension stirred overnight. Treatment of the cooled reaction mixture with water (exotherm) and filtration give a 90–95% yield of tetraphenyltin. Separation and drying of the organic layer from the filtrate, followed by solvent evaporation and distillation, give 65–75% ($\sim$17 g) of allyl(tributyl)tin.

$$(CH_2{=}CHCH_2)_4Sn + 2n\text{-}C_4H_9Li \longrightarrow 2CH_2{=}CHCH_2Li\downarrow + (CH_2{=}CHCH_2)_2Sn(n\text{-}C_4H_9)_2 \qquad (15a)$$

Pure allyllithium can be prepared by placing a solution of *n*-butyllithium in pentane or hexane ($\sim$0.5 *M*) in a Schlenk flask of appropriate size (Fig. 13a–c, septum closure) and slowly injecting into the stirred reagent an amount of tetraallyltin equal to one-half the moles of the lithium reagent. The suspension is decanted through a medium glass frit (Fig. 26), and the solid allyllithium washed with dry, deoxygenated pentane. The Schlenk receiver (Fig. 13c) connected to the filter is changed, the allyllithium dissolved off the frit by dry ether, and the ethereal solution of allyllithium collected in another Schlenk vessel (c′).

Allyllithium is a colorless solid that is pyrophoric in air, sparingly soluble in pentane, and readily soluble in ethyl ether or THF. In the latter two solvents it exhibits degrees of association ranging from 1.4 to 10.[15] At room

[14] If a pale-green or blue color signaling the start of the cleavage is not noticed some time after a small proportion of allyl phenyl ether has been added, a pinch of biphenyl is introduced [see J. J. Eisch, *J. Org. Chem.* **28,** 707 (1963)].

[15] P. West, J. I. Purmont, and S. V. McKinley, *J. Am. Chem. Soc.* **90,** 797 (1968).

temperature an ethereal solution of allyllithium displays an nmr spectrum of the AX_4 type,[16] with the following signals (relative to the central peak of the ether triplet): A, −5.52, q; X, −1.35, d. The shape of the spectrum is temperature-dependent: At −87° it assumes the pattern of the AA′BB′X type.[15] The ultraviolet spectrum shows its long-wavelength maximum at 315 nm (THF),[17] and the infrared spectrum exhibits its C=C bond stretch at 1540 cm^{-1} (Et_2O)[13]. Studies of ^6Li and ^7Li infrared and Raman,[18] ^7Li nmr,[19] and ^{13}C nmr spectra[20] of allyllithium have also been made.

3. *p*-Bromophenyllithium

Organolithium reagents not readily available from the organic halide and lithium metal can often be obtained in high yield by the halogen–lithium exchange reaction.[21] This advantageous procedure is illustrated by the preparation of *p*-bromophenyllithium from *p*-dibromobenzene,[22] with little formation of the dilithium reagent. The preferential monobromo exchange is shown by the high yields of silyl or germyl derivatives obtainable from this reaction.

n-Butyllithium is the usual reagent of choice for such halogen–lithium exchanges, although *sec*-butyllithium and *tert*-butyllithium can be superior in certain situations. The last reagent is particularly suitable for organic iodides[23] in the so-called Trapp medium[24] (a mixture of THF, Et_2O, and pentane in a volume ratio of 3:1:1) at −90 to −110°.

Procedure:

$$p\text{-}BrC_6H_4Br + n\text{-}C_4H_9Li \longrightarrow p\text{-}BrC_6H_4Li + n\text{-}C_4H_9Br \qquad (16)$$

[16] C. S. Johnson, Jr., M. A. Weiner, J. S. Waugh, and D. Seyferth, *J. Am. Chem. Soc.* **83,** 1306 (1961).

[17] R. Waack and M. A. Doran, *J. Phys. Chem.* **68,** 1148 (1964).

[18] C. Sourisseau, B. Pasquier, and J. Hervieu, *Spectrochim. Acta* **31A,** 287 (1975).

[19] P. A. Scherr, R. J. Hogan, and J. P. Oliver, *J. Am. Chem. Soc.* **96,** 6055 (1974).

[20] J. P. C. M. van Dongen, H. W. D. van Dijkman, and M. J. A. de Bie, *Rec. Trav. Chim. Pays-Bas* **93,** 29 (1974).

[21] R. G. Jones and H. Gilman, *in* "Organic Reactions" (R. Adams, ed.), Vol. VI, p. 339, Wiley, New York, 1951.

[22] H. Gilman and R. G. Jones, *J. Am. Chem. Soc.* **63,** 1443 (1941).

[23] H. Neumann and D. Seebach, *Tetrahedron Lett.* 4839 (1976).

[24] G. Köbrich and H. Trapp, *Chem. Ber.* **99,** 680 (1966).

A 1-liter three-necked flask equipped as in Fig. 19 (without drain) is charged with 26.0 g (0.11 mole) of *p*-dibromobenzene and 150 ml of anhydrous ethyl ether. The stirred solution is treated, in one portion, with 0.10 mole of freshly prepared *n*-butyllithium in ether (from *n*-butyl bromide). After 40-min the Gilman color test IIA is negative. Then 26.6 g (0.090 mole) of chloro(triphenyl)silane dissolved in 200 ml of ether is added over a 40-min period. After 2 hr at reflux the suspension is hydrolyzed with dilute aqueous hydrogen chloride. A 23.9-g portion of the *p*-bromophenyl(triphenyl)silane is filtered off, mp 171–172°.[25] The separated and dried ether layer is evaporated, and the residue recrystallized from a benzene–95% ethanol pair to yield 10 g more, mp 170–171°. The total yield is 91%.

A similarly prepared batch of *p*-bromophenyllithium is treated with chloro(triphenyl)germane, mp 110–115°, to yield 83% of *p*-bromophenyl-(triphenyl)germane, glistening needles, mp 158.5–160°.[26]

A study of the proton nmr spectra of several aryllithium derivatives, such as the phenyl-, *m*-toly-, *p*-tolyl-, *m*-chloro-, and *p*-chlorophenyllithium reagents, has been published, and attention is called to the deshielded protons ortho to the carbon–lithium bond.[27]

4. Benzyllithium

Progress toward a useful preparation of benzyllithium reflects the successive impact of newer donor solvents on the whole field of organometallic chemistry. Two time-tested methods in organolithium chemistry, the lithiation of toluene by *n*-butyllithium[28] and the reaction of benzyl halides with lithium,[29] both fail to give other than traces of benzyllithium in ethyl ether solution. In lithiations conducted in ether–THF mixtures at 25°, however, ~25% yields of benzyllithium have been obtained.[30] Finally, by employing 1:1 complexes of *n*-butyllithium with *N*,*N*,*N*′,*N*′-tetramethylethylenediamine (TMEDA) or 1,4-diazabicyclo[2.2.2]octane (DABCO) toluene can be converted almost quantitatively to benzyllithium within minutes at 25°.[31]

The cleavage of benzylic halides to produce benzyllithium fares no better in more Lewis-basic solvents, but other cleavages become possible. Although

[25] H. Gilman and H. W. Melvin, Jr., *J. Am. Chem. Soc.* **72,** 995 (1950).

[26] J. J. Eisch, unpublished studies (1955).

[27] (a) J. Parker and J. A. Ladd, *J. Organometal. Chem.* **19,** 1 (1969); (b) G. Fraenkel, D. A. Adams, and R. R. Dean, *J. Phys. Chem.* **72,** 944 (1968).

[28] J. Mallan and R. L. Bebb, *Chem. Rev.* **69,** 693 (1969).

[29] H. Gilman and R. D. Gorsich, *J. Am. Chem. Soc.* **77,** 3134 (1955).

[30] H. Gilman and B. J. Gaj, *J. Org. Chem.* **28,** 1725 (1963).

[31] (a) G. G. Eberhardt and W. A. Butte, *J. Org. Chem.* **29,** 2928 (1964); (b) A. W. Langer, Jr., *Trans. N.Y. Acad. Sci.* **27,** 741 (1965).

benzyl alkyl ethers are resistant to lithium cleavage in ether, they are readily cleaved in THF at low temperatures to give high yields of benzyllithium.[32]

The metalation of toluene is undoubtedly the superior way to prepare pure solvates of benzyllithium, such as are desirable for structural studies. However, the cleavage of benzyl ethers seems to be superior when large amounts of benzyllithium solution are required.

Procedure A:

$$C_6H_5\text{—}CH_3 + n\text{-}C_4H_9Li\cdot D \xrightarrow{25^\circ} C_6H_5\text{—}CH_2Li\cdot D + n\text{-}C_4H_{10}$$

D = TMEDA, DABCO (17)

In a 100-ml Schlenk flask (Fig. 13a–c, septum closure) are placed 30 ml of toluene and 11.2 g (10 mmoles) of anhydrous DABCO or 11.6 g of TMEDA. Then 10 mmoles of *n*-butyllithium in hexane is injected into the solution. Immediately bright yellow needles of the benzyllithium–amine complex begin to precipitate. Filtration and washing of the product give ~90% yields of benzyllithium. Alternatively, the cooled reaction suspension can be treated with 1.82 g (10 mmoles) of benzophenone in 20 ml of toluene and then hydrolyzed. The dried and evaporated organic layer yields 89% of benzyldiphenylcarbinol.[31]

Procedure B:

$$C_6H_5\text{—}CH_2OCH_2CH_3 + 2Li \xrightarrow[-10^\circ]{THF} C_6H_5\text{—}CH_2Li + LiOCH_2CH_3 \quad (18)$$

In a 200-ml Schlenk flask (Fig. 19, with drain) are placed 60 ml of anhydrous THF and 6.94 g (1 g-atom) of finely and freshly cut lithium metal. With magnetic stirring and cooling at −10° a small amount of a solution of 20.4 g (0.15 mole) of benzyl ethyl ether in 30 ml of anhydrous ethyl ether is added, and the mixture is stirred until coloration signals the start of reaction. Then the balance of the solution is added at a rate of 20 drops/min. After addition the mixture is stirred for a further 60 min at −10°. Double-titration analysis gives yields of 80–85%.[32]

Benzyllithium displays a long-wavelength absorption maximum at 330 nm (THF)[17] and nmr absorptions for its ortho, meta, and para protons at (δ) 6.09, 6.30, and 5.50 ppm, respectively (THF).[33] The crystal structure of the 1:1 complex of benzyllithium with triethylenediamine has been

---

[32] (a) H. Gilman and H. A. McNinch, *J. Org. Chem.* **26,** 3723 (1961); (b) H. Gilman and G. L. Schwebke, *ibid.* **27,** 4259 (1962).

[33] V. R. Sandel and H. H. Freedman, *J. Am. Chem. Soc.* **85,** 2328 (1963).

determined,[34] and the ^{7}Li and ^{13}C nmr spectra have been compared with those of other organolithium compounds.[19,20]

5. Phenoxymethyllithium

The rapidity of tin–lithium exchange processes, even at low temperatures, permits the preparation of organolithium reagents too unstable to persist at room temperature. Such a compound is phenoxymethyllithium, which can be generated and chemically trapped at $-78°$ but decomposes at higher temperatures. Upon warming to room temperature and working up by hydrolysis, little anisole (the expected protonolysis product) is detected, hence an α-elimination leading to methylene can be presumed. In a similar manner, the lithiation of benzyl phenyl ether in ethyl ether solution leads preponderantly to the elimination of lithium phenoxide and the generation of a phenylmethylene intermediate.[35]

Procedure:

$$(n\text{-}C_4H_9)_3Sn\text{—}CH_2OC_6H_5 + n\text{-}C_4H_9Li \xrightarrow[-78°]{THF} C_6H_5OCH_2Li + (n\text{-}C_4H_9)_4Sn \quad (19)$$

$$C_6H_5OCH_2Li \xrightarrow[2.\ H_2O]{1.\ (C_6H_5)_2CO} (C_6H_5)_2C(OH)CH_2OC_6H_5 \quad (20)$$

In a 100-ml Schlenk flask (Fig. 13c, septum closure) are placed 3.97 g (10 mmoles) of tri-*n*-butyl(phenoxymethyl)tin and 40 ml of anhydrous THF. With stirring and cooling at $-78°$ the solution is treated with 6.9 ml of 1.6 *M* *n*-butyllithium (11 mmoles) in hexane for 5 min. After 10 min 2.0 g (11 mmoles) of solid benzophenone is added. The reaction solution is allowed to come to room temperature and treated with an aqueous ammonium chloride solution. Separation of the organic layer, drying over anhydrous sodium sulfate, evaporation of solvent, and chromatography of the residue on silica gel with a hexane–ether gradient give 3.5 g (100%) of tetra-*n*-butyltin and product admixed with recovered benzophenone. Recrystallization from hexane gives 1.42 g (49%) of 2-phenoxy-1,1-diphenylethanol, mp 99–100°.[36]

6. Diphenylmethyllithium

The two principal routes to this compound involve the abstraction of a proton from diphenylmethane by lithium reagents[31,37] and the cleavage of

---

[34] S. P. Patterman, I. L. Karle, and G. D. Stucky, *J. Am. Chem. Soc.* **92,** 1150 (1970).

[35] U. Schöllkopf and M. Eisert, *Justus Liebigs Ann. Chem.* **664,** 76 (1963).

[36] J. J. Eisch and J. E. Galle, unpublished studies (1978).

[37] (a) H. Gilman and R. L. Bebb, *J. Am. Chem. Soc.* **61,** 109 (1939); (b) J. J. Eisch and W. C. Kaska, *J. Org. Chem.* **27,** 3745 (1962).

sym-tetraphenylethane by lithium metal.[30,38] Both reactions are accelerated by using THF in place of ethyl ether. Time and temperature are essential in preserving the reagent: Prolonged reaction at 65° leads to proton abstraction and to cleavage of THF.[38] Reaction of diphenylmethane with lithium metal, even in a dispersed form, is very slow; the dispersion requires 5 days for a 60% yield of diphenyllithium.[30] The use of 15% biphenyl as a solubilizing agent for lithium metal permits a 50% yield within 24 hr.[37b] Alternatively, diphenylmethane can be smoothly lithiated by the *n*-butyllithium–TMEDA complex in the manner described in Section B,4.[31]

The cleavage of *sym*-tetraphenylethane by lithium metal proceeds rapidly at 25° if small amounts of biphenyl are present as a metal-solubilizing agent. The high yields of diphenylmethyllithium obtained recommend this method if the presence of biphenyl is compatible with further uses of the reagent.[38]

Procedure:

$$(C_6H_5)_2CH{-}CH(C_6H_5)_2 + 2Li \xrightarrow[25^\circ]{(C_6H_5)_2,\ THF} 2(C_6H_5)_2CHLi \qquad (21)$$

In a 150-ml Schlenk flask (Fig. 13c, with stopper) are placed 50 ml of anhydrous THF, 6.7 g (20 mmoles) of *sym*-tetraphenylethane, and 1.0 g (7 mmoles) of biphenyl. Then 305 mg (44 mg-atoms) of lithium wire is cut finely and dropped into the flask. The suspension is stirred vigorously at 20–25° for 2–3 hr. Analysis by treating the filtered solution at −78° with solid carbon dioxide gives yields of 75–85% of diphenylacetic acid.

Diphenylmethyllithium displays a long-wavelength absorption maximum at 496 nm in THF and 434 nm in ethyl ether,[39] and nmr absorption for its ortho, meta, and para protons at (δ, THF) 6.51, 6.54, and 5.65 ppm, respectively.[33]

7. Triphenylmethyllithium

Various lithium reagents remove a proton efficiently from triphenylmethane. For structural studies amine complexes of alkyllithium compounds are the reagents of choice. A 1:1 complex of *n*-butyllithium and TMEDA readily yields the glistening red needles of the triphenylmethyllithium–TMEDA complex.[40] Although lithium dispersions in THF can form triphenylmethyllithium from triphenylmethane, the reaction is erratic and requires days.[30] Almost quantitative yields are obtained by using biphenyl

---

[38] From *sym*-tetraphenylethane and lithium metal (dispersion or its 2:1 biphenyl adduct) at 65° in THF, carbonation gave some diphenylacetic acid, diphenylmethane, and 5,5-diphenyl-1-pentanol: J. J. Eisch, *J. Org. Chem.* **28,** 707 (1963).

[39] R. Waack and M. A. Doran, *J. Am. Chem. Soc.* **85,** 1651 (1963).

[40] J. J. Brooks and G. D. Stucky, *J. Am. Chem. Soc.* **94,** 7346 (1972).

as a solubilizing agent for the lithium metal. In this case, however, the biphenyl also acts as a hydrogen sink, being reduced to phenylcyclohexane.[37b]

Procedure A:

$$6(C_6H_5)_3CH + 6Li + (C_6H_5)_2 \xrightarrow{THF} 6(C_6H_5)_3CLi + \langle S \rangle\!-\!C_6H_5 \qquad (22)$$

In a 150-ml Schlenk flask (Fig. 19) equipped with a reflux condenser are placed 5.35 g (22 mmoles) of triphenylmethane, 570 mg (3.7 mmoles) of biphenyl, and 75 ml of anhydrous THF. The solution is treated with 150 mg (23 mg-atoms) of finely and freshly cut lithium wire. The suspension, which soon turns red, is stirred vigorously for 6 hr at 20–25° and 3 hr at 66°. Treatment of the triphenylmethyllithium solution with solid carbon dioxide gives 85–95% yields of triphenylacetic acid.[37b]

Procedure B:

$$(C_6H_5)_3CH + n\text{-}C_4H_9Li \xrightarrow{THF-Et_2O} (C_6H_5)_3CLi + n\text{-}C_4H_{10} \qquad (23)$$

In a 150-ml Schlenk flask (Fig. 13c) equipped with a septum are placed 5.0 g (20 mmoles) of triphenylmethane and 50 ml of anhydrous THF. Then a solution of 30 ml of 1.0 *M* *n*-butyllithium in ethyl ether is injected. After 4 hr at 0° carbonation gives a 75–85% yield of triphenylacetic acid.[39]

Triphenylmethyllithium displays a long-wavelength absorption maximum at 500 nm (THF)[39] and nmr absorptions for its ortho, meta, and para protons at (δ, THF) 6.09, 6.30, and 5.50 ppm, respectively.[33]

8. (*E*,*E*)-1,2,3,4-Tetraphenyl-1,3-butadien-1,4-ylidenedilithium

This preparation illustrates an important route to dimetallo derivatives and, at the same time, produces a key intermediate for the synthesis of many metallocyclic derivatives. Being a widely used reaction, the formation of this 1,4-dilithio-1,2,3,4-tetraphenylbutadiene has been described by many workers, and the consensus is that the reaction can often give erratic results.[41–44] The fault lies in the complexity of the reaction of lithium with diphenylacetylene, which can lead, depending upon the solvent and the

[41] L. I. Smith and H. H. Hoehn, *J. Am. Chem. Soc.* **63,** 1184 (1941).
[42] F. C. Leavitt, T. A. Manuel, F. Johnson, L. U. Matternas, and D. S. Lehman, *J. Am. Chem. Soc.* **82,** 5099 (1960).
[43] E. H. Braye, W. Hübel, and I. Caplier, *J. Am. Chem. Soc.* **83,** 4406 (1961).
[44] H. Gilman, S. G. Cottis, and W. H. Atwell, *J. Am. Chem. Soc.* **86,** 1596 (1964).

temperature, to the mono-[45] or cis-bis lithio adducts:[46]

$$C_6H_5-\overset{Li}{C}=\overset{Li}{C}-C_6H_5 \xleftarrow{2Li} C_6H_5-C\equiv C-C_6H_5 \xrightarrow{Li} C_6H_5-\overset{Li}{C}=\dot{C}-C_6H_5 \quad (24)$$

The radical couples with itself to give the (E,E) dimer:[45]

$$2C_6H_5-\overset{Li}{C}=\dot{C}-C_6H_5 \longrightarrow (C_6H_5)(Li)C=C(C_6H_5)-C(C_6H_5)=C(Li)(C_6H_5) \quad (25)$$

With prolonged reactions, the (E,E) dimer isomerizes (catalyzed by the lithium or a strong donor like THF) and undergoes ring closure to 2,3,4-triphenyl-1-naphthyllithium[47]:

$$(C_6H_5)(Li)C=C(C_6H_5)-C(C_6H_5)=C(Li)(C_6H_5) \xrightarrow{\Omega} Li(C_6H_5)C=C(C_6H_5)-C(C_6H_5)=C(Li)(C_6H_5) \longrightarrow \text{1-Li-2,3,4-}(C_6H_5)_3\text{-naphthalene} \quad (26)$$

The principal experimental difficulty in this reaction therefore is in maximizing the amount of the desired (E,E) dimer and knowing its yield. Without a reliable assay, improper amounts of other reagents (e.g., R_2SiCl_2 or R_2SnCl_2) will be used in subsequent steps. As is seen in the preparation of 1,1-dimethyl-2,3,4,5-tetraphenylstannole (Section F,21), such incorrect ratios of reactants can cause the synthesis to fail.

Several strategems can give an estimate of the amount of (E,E) dimer present: (*a*) The unreacted lithium can be filtered off or removed with long forceps, and thereby the lithium consumed can be calculated; (*b*) an aliquot can be hydrolyzed, and the (*E*,*E*)-1,2,3,4-tetraphenyl-1,3-butadiene determined by column or gas chromatography; or (*c*) the reaction product can be dissolved in THF, an aliquot hydrolyzed and titrated with acid, and the

---

[45] D. A. Dodley and A. G. Evans, *J. Chem. Soc.* 418 (1967); 107 (1968); (b) R. E. Sioda, D. O. Cowan, and W. S. Koski, *J. Am. Chem. Soc.* **89,** 230 (1967).

[46] G. Levin, J. Jagur-Grodzinski, and M. Swarc, *J. Am. Chem. Soc.* **92,** 2268 (1970).

[47] (a) W. Schlenk and E. Bergmann, *Justus Liebigs Ann. Chem.* **463,** 1 (1928); (b) E. Bergmann and O. Zwecker, *Ibid.* **487,** 155 (1931).

assumption then made that soluble lithium exists principally as the (E,E) dimer. A final method, described below, is to isolate the dimer and determine its effective molecular weight, hence its ether content, by hydrolysis of an aliquot.[48]

Procedure:

$$2Li + 2C_6H_5{-}C{\equiv}C{-}C_6H_5 \xrightarrow{Et_2O} (C_6H_5)(Li)C{=}C(C_6H_5){-}C(C_6H_5){=}C(Li)(C_6H_5) \qquad (27)$$

In a 100-ml Schlenk vessel (Fig. 13c) are placed 8.9 g (50 mmoles) of pure diphenylacetylene (free of *trans*-stilbene), 40 ml of anhydrous ethyl ether, and 227 mg of lithium metal (previously hammered into foil). The mixture is magnetically stirred for 1.75 hr, and the remaining lithium then removed with forceps. The solution is then stirred an additional 15 hr to remove any remaining specks of lithium. The resulting yellow precipitate is filtered off on a coarse frit under nitrogen (Fig. 26), washed with dry, deoxygenated ether to remove red impurities, and dried *in vacuo*. The dried (*E,E*)-1,2,3,4-tetraphenyl-1,3-butadiene-1,4-ylidenedilithium amounts to 6–9 g (50–70%). Hydrolysis and acid titration of an aliquot gives an apparent equivalent weight of 250; hence the product exists principally as a dietherate (MW 518). Under nitrogen the solid is relatively stable for months at room temperature. Treatment of a 4-month-old sample with pure, dry THF and deuterium oxide gave the tetraphenylbutadiene which, after recrystallization, was shown by nmr spectroscopy to have 1.70 deuterons at the 1 and 4 positions.

[48] J. J. Eisch and J. E. Galle, unpublished studies (1976).

C. Compounds of Group IIA

1. General Preparative and Analytical Methods

Among the organometallic derivatives of beryllium, magnesium, calcium, strontium, and barium, those of magnesium have maintained their pre-eminence in synthesis for fourscore years. The relative scarcity and high toxicity of beryllium remain significant barriers to developing the chemistry of its organometallic compounds. The derivatives of calcium, strontium, and barium have received only sporadic attention,[49] but recently refined preparative methods may signal promising developments, especially for the potentially useful organocalcium reagents.[50]

The ease of preparing organomagnesium compounds from diverse organic halides, their ready solubility and great storability in ethereal solvents,[51] and their high reactivity toward polar organic functional groups (C=O, C=N, C=S, C—X, etc.) and salts of less electropositive metals [Eq. (1)], make them superb reagents in chemical synthesis.[52]

The most commonly applicable preparative method is the interaction of an organic halide with magnesium metal. The comments in Part I, Section B, on the role of the metal surface, solvent, and promoters may be useful in conducting a Grignard preparation with an unstudied or unusual halide. For prompt initiation the reagents and solvent should be made (and kept) scrupulously dry, the magnesium surface scratched mechanically (by stirring) or etched chemically (iodine or ethylene bromide), and the reactants maintained under dry nitrogen. For less reactive halides, such as chlorides, or the formation of less soluble Grignard reagents, such as allylmagnesium

---

[49] G. Bähr and H. O. Kalinowski, *in* "Methoden der Organischen Chemie" (E. Müller, ed.), Vol. XIII/2a, Organometallic Compounds, pp. 529–552. Georg Thieme Verlag, Stuttgart, Germany, 1973.

[50] (a) M. A. Zemlyanichenko, N. I. Sheverdina, I. M. Viktorova, N. P. Barminova, and K. A. Kocheshkov, *Dokl. Akad. Nauk. SSSR* **194,** 95 (1970); (b) N. Kawabata, A. Matsumura, and S. Yamashita, *Tetrahedron* **29,** 1069 (1973).

[51] V. Grignard [*C. R. Acad. Sci. Paris* **151,** 322 (1910)] pointed out early in his work that diethyl ether was attacked very slowly by these reagents even at $>180°$.

[52] (a) M. S. Kharasch and O. Reinmuth, "Grignard Reactions of Nonmetallic Substances," Prentice Hall, Englewood Cliffs, New Jersey, 1954; (b) K. Nützel, *in* "Methoden der Organischen Chemie" (E. Müller, ed.), Vol. XIII/2a, Organometallic Compounds, pp. 47–528. Georg Thieme Verlag, Stuttgart, Germany, 1973.

chloride, THF is superior to ethyl ether.[53] Preparation of Grignard reagents can also be achieved in hydrocarbon media, either in suspension[54] or as soluble complexes of triethylamine[55]:

$$2RX + 2Mg \xrightarrow{\text{donor}} 2RMgX \rightleftharpoons R_2Mg + MgX_2 \tag{28}$$

The other important routes to organomagnesium reagents are actually interconversion reactions of other magnesium compounds, such as (*a*) the metalation of carbon acids, such as terminal alkynes, cyclopentadienes, arylmethanes, and methyl ketones or imines by alkylmagnesium reagents (see the preparation of trimethyl(phenylethynyl)silane and -tin [Sections F,3 and F,13, and Eq. (29)]; (*b*) the isolation of halogen-free dialkylmagnesium by shifting the Schlenk equilibrium through a precipitation of magnesium halide with dioxane [Eq. (30)]; (*c*) the conversion of an organolithium reagent with magnesium halide [Eq. (31)]; (*d*) the titanium-catalyzed transfer of magnesium hydride from an alkyl Grignard reagent to an olefinic substrate [Eq. (32)]; and (*e*) the uncatalyzed or nickel-catalyzed addition of reactive Grignard reagents to olefinic or acetylenic alcohols[56,57] or amines[58] [Eq. (33)]:

(29) [structures: R—Mg—X with cyclopentenoquinoline → MgX-substituted cyclopentenoquinoline (ref. 59); R—Mg—X with cyclopentadiene → cyclopentadienyl-MgX (ref. 60)]

$$2RMgX \xrightarrow{C_4H_8O_2} R_2Mg + MgX_2\,(\text{dioxane})\downarrow \tag{30}$$

$$RLi + MgX_2 \longrightarrow RMgX + LiX \tag{31}$$

$$R{-}CH(R){-}CH_2MgX + R'{-}CH{=}CH_2 \xrightarrow{Ti(IV)} R'{-}CH_2{-}CH_2MgX + R_2C{=}CH_2 \tag{32}$$

$$R{-}C(OH)(R){-}(CH_2)_nCH{=}CH_2 + 2R'MgX \xrightarrow{Ni} R{-}C(R)(O{-}Mg{-})(CH_2)_n{-}CH{-}CH_2R' \tag{33}$$

---

[53] (a) H. Normant, *Bull. Soc. Chim. Fr.* 1434 (1963); (b) H. E. Ramsden, A. E. Balint, W. R. Whitford, J. J. Walburn, and R. Cserr, *J. Org. Chem.* **22,** 1202 (1957).
[54] D. Bryce-Smith and G. F. Cox, *J. Chem. Soc.* 1050 (1956); 1175 (1961).
[55] E. C. Ashby and R. Reed, *J. Org. Chem.* **31,** 971 (1966).
[56] J. J. Eisch, J. H. Merkley, and J. E. Galle, *J. Org. Chem.* **44,** 587 (1979).
[57] J. J. Eisch and J. H. Merkley, *J. Am. Chem. Soc.* **101,** 1148 (1979).
[58] H. G. Richey, W. F. Erickson, and A. S. Heyn, *Tetrahedron Lett.* 2183, 2187 (1971).
[59] J. J. Eisch and F. J. Gadek, *J. Org. Chem.* **36,** 3376 (1971).
[60] C. Courtot, *Ann. Chim.* [9], **4,** 58 (1915).

The qualitative detection of organomagnesium compounds in solution is best performed by means of Gilman's color test I (Part I, Section B,9) or by derivatizing the organometallic with carbon dioxide or benzophenone (for solid products) or with methyl iodide or chlorotrimethylsilane (for volatile products). For quantitative determination, simple hydrolysis of a solution aliquot and a single titration of the resulting alkali with standard acid may give a useful estimate of the carbon–magnesium bonds if no other source of magnesium hydroxide is present. Obviously this method is unsuitable for alkylmagnesium alkoxides. A sharper end point in a titration with phenolphthalein indicator is obtained if the organomagnesium aliquot is hydrolyzed with a known excess of standard acid and the resulting hydrolysate back-titrated with standard sodium hydroxide (see Sections B,9 and D,4 for applicable technique[61]).

Alternatively, the titration of organomagnesium compounds in ethereal solution with 1-butanol in the presence of 1,10-phenanthroline has been recommended as a very simple and reliable analytical method.[62]

Yet, for a direct estimate of the actual carbon–magnesium bonds in an aliquot, some chemical derivatizing process must be invoked. Methanolysis of arylmagnesium compounds coupled with a gas chromatographic estimation of the resulting arenes,[63] and spectrophotometric determination of the benzophenone consumed by Grignard samples,[64] illustrate such useful direct analyses.

2. α-Deuteriovinylmagnesium Chloride

The broad importance of vinylmetallic compounds, both in structural and in mechanistic studies (Section F,9), often creates a need for specifically deuterated derivatives. For example, in a recent study on the reaction of α,β-epoxyalkylsilanes with organolithium reagents, the α-deuterio derivative was required to distinguish between paths A and B[65]:

$$CH_2{=}CHR \xleftarrow[\substack{-R_4Si \\ 2.\ H{-}S(THF)}]{1.\ 2RLi} \underset{\substack{| \\ Li}}{\overset{O}{CH_2{-}C{-}SiR_3}} \xleftarrow[\substack{-RD \\ \text{path A}}]{RLi} \underset{\substack{| \\ D}}{\overset{O}{CH_2{-}C{-}SiR_3}} \xrightarrow[-R_4Si]{RLi} \text{(path B)} \tag{34}$$

$$CH_2{=}CDR \xleftarrow[-Li_2O]{RLi} \underset{\substack{| \\ D}}{\overset{O}{CH_2{-}C{-}Li}}$$

[61] H. Gilman, E. A. Zoellner, and J. B. Dickey, *J. Am. Chem. Soc.* **51,** 1576 (1929).

[62] S. C. Watson and J. F. Eastham, *J. Organometal. Chem.* **9,** 165 (1967).

[63] L. V. Guild, C. A. Hollingsworth, D. H. McDaniel, and J. H. Wotiz, *Anal. Chem.* **33,** 1156 (1961).

[64] R. D'Hollander and M. Anteunis, *Bull. Soc. Chim. Belges* **72,** 77 (1963).

[65] J. J. Eisch and J. E. Galle, *J. Org. Chem.* **41,** 2615 (1976).

Since the epoxide is readily prepared from the vinylsilane, synthesis of the latter requires the α-deuterated vinyl Grignard reagent:

$$R_3SiCl + CH_2{=}CDMgCl \xrightarrow{-MgCl_2} CH_2{=}CDSiR_3 \xrightarrow{ArCO_3H} \overset{O}{CH_2{-}CDSiR_3} \quad (35)$$

The following procedure is a convenient way of generating and drying the vinyl chloride and of generating the Grignard reagent in THF.[66]

Procedure:

$$CH_2{=}CCl_2 + DBr \xrightarrow{h\nu} BrCH_2{-}CCl_2{-}D \xrightarrow[-ZnBrCl]{Zn} H_2C{=}C(D)Cl \xrightarrow[THF]{Mg} H_2C{=}C(D)MgCl \quad (36)$$

In a quartz reaction vessel are placed 60 ml (72.7 g, 0.75 mole) of vinylidene chloride. While the contents are irradiated at 254 nm in a Rayonet reactor, Model PRP-100, equipped with low-pressure mercury lamps, a rapid current of deuterium bromide (generated by the slow addition of D_2O to PBr_3) is bubbled into the liquid. When no more deuterium bromide is absorbed (18 hr), the liquid is diluted with methylene chloride and then washed with 5% aqueous sodium bicarbonate solution. After drying over anhydrous potassium carbonate, the solvent is evaporated and the 1-bromo-2,2-dichloroethane-2-*d* is distilled, bp 133–135°. The yield ranges up to 60%, depending upon the efficiency with which the deuterium bromide is introduced. If the deuterium bromide is added too slowly, much telomerization of the vinylidene chloride occurs.

A 100-ml three-necked flask (Fig. 19, without drain) is charged with 10 g (0.153 g-atom) of granulated zinc and 40 ml of 95% ethanol. The condenser is attached to a trap cooled at −78°. Then 11.5 g (64 mmoles) of the deuterated haloethane is added dropwise to the stirred and refluxing suspension. The vinyl chloride is caught in the cold trap and then, by warming, volatilized through successive drying tubes containing granular phosphorus pentoxide and calcium hydride, respectively.

The dried vinyl chloride is passed into a 150-ml three-necked flask (Fig. 19) containing a stirred suspension of 1.2 g (49 mg-atoms) of magnesium turnings and 25 ml of anhydrous THF. After the Grignard reagent has been formed, a solution of 10.5 g (36 mmoles) of chloro(triphenyl)silane in 50 ml of THF is added and the mixture heated at reflux for 24 hr. Hydrolysis with aqueous ammonium chloride, separation of the organic layer and drying ($MgSO_4$), and evaporation of the solvent give a crude product which is

[66] J. E. Francis and L. C. Leitch, *Can. J. Chem.* **35,** 500 (1957).

chromatographed on Fluorosil with a hexane eluent. The triphenyl(vinyl)-silane obtained is recrystallized from a methylene chloride–95% ethanol pair to give 8.0 g. This constitutes a 76% yield.

3. Allylmagnesium Bromide

The preparation of allylic or benzylic Grignard reagents can present special problems. With bromides or iodides, usual procedures lead to low yields of the Grignard reagent because of the ease with which the generated magnesium compound couples with R—X to form R—R. Coupling can be minimized by slow introduction of the halide into a liberal excess of magnesium.[67] The lesser tendency of the chloride to undergo coupling makes this procedure unnecessary in preparing allylmagnesium chloride. However, since this Grignard reagent is insoluble in ether but soluble in THF, the latter medium is superior for its preparation.[68]

Procedure:

$$CH_2{=}CHCH_2Br + Mg \xrightarrow{Et_2O} CH_2{=}CHCH_2MgBr \qquad (37)$$

Caution: Many allylic and benzylic halides are lachrymators and skin irritants. Disposable gloves should be worn, and transfers carried out in an efficient hood.

In a 1-liter three-necked flask equipped as in Fig. 19 are placed 48.9 g (2.02 g-atoms) of magnesium turnings and 150 ml of dry ethyl ether. After initiation, the balance of a solution of 40.8 g (0.377 mole) of freshly distilled allyl bromide in 190 ml of ether is added dropwise to the vigorously stirred solution over a 2-hr period. After the solution has cooled, it is drained into a calibrated addition funnel previously flushed with nitrogen. Acid titration indicates a yield of 80–85%. The residual magnesium can be washed with dry ether and maintained under nitrogen for a subsequent preparation.

As a derivatizing assay, 10 mmoles of this Grignard solution can be treated, under nitrogen, with 2.6 g (10.1 mmoles) of benzophenone anil dissolved in 20 ml of dry ether. After hydrolysis with aqueous ammonium chloride solution, the organic residue from the separated, dried, and evaporated ether layer is essentially a quantitative yield of α-allylbenzhydrylaniline. Recrystallization from petroleum ether gives >90% pure product, mp 77–79°. For unknown Grignard concentrations, an excess of the anil is used and the product separated by column chromatography on alumina.[57]

The nmr spectrum of allylmagnesium bromide in ethyl ether solution

---

[67] H. Gilman and J. H. McGlumphy, *Bull. Soc. Chim. Fr.* **43,** 1325 (1928).

[68] (a) M. S. Kharasch and C. F. Fuchs, *J. Org. Chem.* **9,** 359 (1944); (b) H. Normant, *C. R. Acad. Sci. Paris* **239,** 1510 (1954).

displays an AX_4 pattern, with the four identical methylene protons coupled to the β proton with $J = 12$ Hz.[69] The infrared spectrum in ether has an absorption at 1588 cm^{-1}, which is assigned to the C=C stretch; in THF this band is shifted to 1565 cm^{-1}.[70]

4. Diallylmagnesium

The preparation of halogen-free but ether-solvated magnesium alkyls is often advantageous in reactions where the halogen may influence the reactivity, reaction pathways, or solubility of the organomagnesium reagent. The use of dialkylmagnesium reagents in cleaving epoxides without skeletal rearrangement[71] or in carbomagnesiating unsaturated alcohols[56,57] illustrates their preparative advantages.

Such diorganomagnesium reagents have been obtained from many Grignard solutions by upsetting the classic Schlenk equilibrium [Eqs. (28) and (30)] through precipitating MgX_2 (and some RMgX) as a solvate of dioxane.[72]

Procedure:

$$2CH_2{=}CHCH_2MgBr \rightleftharpoons (CH_2{=}CHCH_2)_2Mg + MgBr_2 \xrightarrow{C_4H_8O_2} (CH_2{=}CHCH_2)_2Mg + CH_2{=}CHCH_2MgBr \cdot C_4H_8O_2 \downarrow + MgBr_2 \cdot C_4H_8O_2 \downarrow \quad (38)$$

A solution of 300 ml of allylmagnesium bromide in ether (0.10 mole, prepared as described in Section F,19, contained in a 500-ml Schlenk flask (Fig. 13c) is treated, while being stirred, with 35 ml (0.41 mole) of purified 1,4-dioxane freshly distilled from calcium hydride.[73] With a slight exotherm a fine white precipitate is formed. The mixture is stirred for 12 hr and then allowed to settle. By siphon or syringe the supernatant liquid is removed, and the precipitate washed with 100 ml of a 90:10 ether–dioxane mixture. The combined organic layer gives a negative test for halide and contains 35 mmoles of diallylmagnesium (lit.[72] 33 mmoles).

For a derivatizing assay, ~0.03 mole of this reagent can be treated with 0.06 mole of 9-fluorenone. After hydrolysis with ammonium chloride solution, the organic residue from the ether layer is chromatographed on an

---

[69] J. E. Nordlander, W. G. Young, and J. D. Roberts, *J. Am. Chem. Soc.* **83,** 499 (1961).

[70] M. Andrae *et al.*, *Bull. Soc. Chim. Fr.* 1385 (1963).

[71] P. D. Bartlett and C. M. Berry, *J. Am. Chem. Soc.* **56,** 2683 (1934).

[72] (a) W. Schlenk, Jr., *Ber.* **64,** 734 (1931); (b) C. R. Noller and F. B. Hilmer, *J. Am. Chem. Soc.* **54,** 2503 (1932); (c) G. O. Johnson and H. Adkins, *ibid.* **54,** 1943 (1932).

[73] 1,4-Dioxane may be purified by the method of N. A. Milas [*J. Am. Chem. Soc.* **53,** 221 (1931)], distilled from sodium under nitrogen, and kept under nitrogen. Any sample should be tested for peroxides before use (Part I, Section A,3).

alumina column. Recrystallization of the alcohol-containing fractions from cyclohexane gives 92–95% of 9-allyl-9-fluorenol, mp 118–120°.

5. 3-(1-Bromomagnesiooxy-1-cyclohexyl)propylmagnesium Bromide

A serious limitation to the use of Grignard reagents in synthesis has been the necessity of preparing them from organic halides and the metal. If reactive functional groups, such as hydroxyl, were present in the halide, then some kind of chemical masking, such as tetrahydropyranyl ether formation, had to be carried out prior to the Grignard reaction. The recent finding that the magnesium hydride transfer between Grignard reagents and olefins can be catalyzed by titanium salts presents a novel, and potentially most versatile, route to Grignard reagents. Not only can terminal olefins[74] now serve as suitable sources of Grignard reagents but, most significantly, olefinic alcohols and other functionalized alkenes[75] can be used directly, without masking. By simply using two equivalents of the accessible ethyl Grignard reagent, both hydroxyl salt formation and magnesium hydride transfer can be smoothly achieved:

$$HO{-}(CH_2)_n{-}CH{=}CH_2 \xrightarrow[-C_2H_6]{CH_3CH_2MgBr} BrMgO(CH_2)_n{-}CH{=}CH_2 \xrightarrow[Ti^{4+},\ -C_2H_4]{CH_3CH_2MgBr} BrMgO(CH_2)_{n+2}MgBr \quad (39)$$

The cessation of ethylene evolution serves as a convenient monitor of the hydromagnesiation reaction.

Procedure:

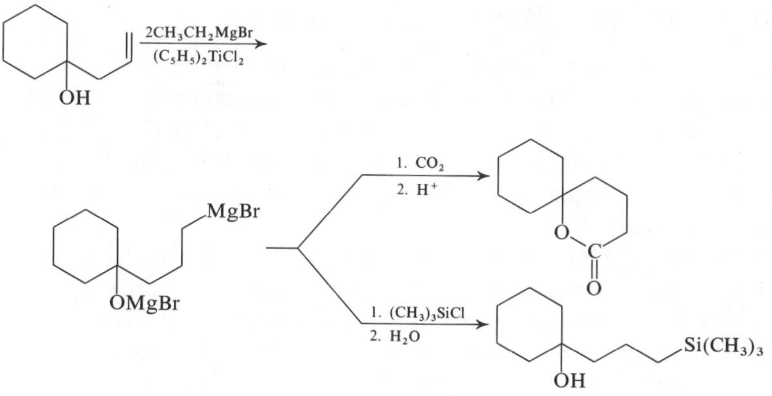

1-allylcyclohexanol (OH) $\xrightarrow[(C_5H_5)_2TiCl_2]{2CH_3CH_2MgBr}$ 1-(3-bromomagnesiopropyl)cyclohexyl–OMgBr

→ 1. CO_2; 2. H^+ → spiro lactone (O–C(=O)) (40)

→ 1. $(CH_3)_3SiCl$; 2. H_2O → 1-(3-trimethylsilylpropyl)cyclohexanol, OH, $Si(CH_3)_3$ (41)

---

[74] (a) G. D. Cooper and H. L. Finkbeiner, *J. Org. Chem.* **27,** 1493 (1962); (b) H. L. Finkbeiner and G. D. Cooper, *Ibid.* **27,** 3395 (1962).

[75] J. J. Eisch and J. E. Galle, *J. Organometal. Chem.* **160,** C8 (1978).

In a 500-ml three-necked flask equipped as in Fig. 19 are placed 14.0 g (0.10 mole) of 1-allylcyclohexanol, 50 mg (2.0 mmoles) of titanocene dichloride, and 200 ml of anhydrous THF. With stirring and cooling in an ice bath the solution is treated with 104 ml of 2.4 *M* ethylmagnesium bromide in ether (0.250 mole) over a period of 15 min. The solution is heated at reflux for 15–20 hr (until ethylene evolution has ceased).

Workup A [Eq. (40)]: The solution is slowly added to a slurry of powdered solid carbon dioxide in THF. The thawed mixture is carefully treated with 300 ml of 2 *N* aqueous sulfuric acid, and the organic layer is then separated, washed with saturated aqueous sodium chloride, and dried over anhydrous magnesium sulfate. The solvent is removed, and the residue heated with 200 ml of benzene in a Dean–Stark trap for 2 hr. The benzene is washed with three 40 ml portions of 5% aqueous sodium hydroxide, dried over anhydrous magnesium sulfate, and distilled to give the lactone, bp 145–150°, which crystallizes upon standing. Acidification of the base extracts, ether extraction, and drying give additional lactone upon distillation (total lactone, 9.2 g, 55%).

Workup B [Eq. (41)]: A solution of 38 g (0.35 mole) of chloro(trimethyl)-silane in 25 ml of anhydrous THF is added dropwise, and the reaction mixture heated at reflux for 2 hr. Hydrolysis with aqueous ammonium chloride solution, drying of the organic layer over anhydrous sodium sulfate, and distillation yield 60–75% 1-(3-trimethylsilyl-1-propyl)cyclohexanol, bp 58–60° at 0.1 mm Hg.

6. (*E*)-2-Phenyl-1-phenylsulfonylvinyllithium and (*E*)-2-Phenyl-1-phenylsulfonylvinylmagnesium Bromide

The following procedure illustrates several points of interest: (*a*) the accelerating action of THF on the proton–lithium exchange reaction, compared with diethyl ether[76]; (*b*) the locoselectivity of proton abstraction[77]; and (*c*) the advantage, in certain cases, of converting organolithium compounds into organomagnesium compounds in order to obtain higher yields in organic reactions.[77]

First, the use of THF permits vinyl sulfones to be metalated rapidly at a remarkably low temperature ($< -90°$) with methyllithium, one of the less reactive alkyllithium reagents. This reagent has the practical advantage of showing the progress of its reaction by the evolution of methane.

Second, a vinyl sulfone such as phenyl *trans*-1-propenyl sulfone has four

[76] (a) H. Gilman and R. D. Gorsich, *J. Org. Chem.* **22,** 687 (1957); (b) H. Gilman and S. Gray, *ibid.* **23,** 1476 (1958).

[77] J. J. Eisch and J. E. Galle, *J. Org. Chem.* **44,** 3279 (1979).

possible sites at which lithiation can take place, the ortho-phenyl (1), the α-vinyl (2), the β-vinyl (3), and the γ-allylic (4) protons:

$$\underset{(4)\,CH_3}{\overset{H^{(3)}}{}}\!\!>C{=}C<\!\!\underset{\underset{(2)}{H}}{\overset{SO_2-C_6H_5\,(1)}{}}$$

Yet at −95°, the only discernible site of attack is the α-vinyl hydrogen.

Third, although organolithium reagents often give cleaner, higher yielding reactions with organic substrates than Grignard reagents, this is not always true. Sometimes the greater reactivity of the lithium derivative leads to undesired side reactions. With enolizable substrates, such as cyclohexanone, the lithium reagent may abstract a proton instead of adding to the carbonyl. By the simple expedient of adding magnesium bromide, the lithium reagent can be converted to the Grignard reagent, which reacts more cleanly with cyclohexanone to yield the desired adduct.

Procedure:

$$(H)(C_6H_5)C{=}C(SO_2Ph)(H) \xrightarrow[\text{THF, }-95^\circ]{CH_3Li} (H)(C_6H_5)C{=}C(SO_2Ph)(Li) + CH_4$$

$$(H)(C_6H_5)C{=}C(SO_2Ph)(Li) \xrightarrow[\text{2. } R_2C{=}O]{\text{1. } MgBr_2} (H)(C_6H_5)C{=}C(SO_2Ph)C(R)(R)OH \quad (42)$$

$$(H)(C_6H_5)C{=}C(SO_2Ph)(Li) \xrightarrow{CH_3I} (H)(C_6H_5)C{=}C(SO_2Ph)(CH_3) \quad (43)$$

In a 150-ml Schlenk flask (Fig. 13c, septum closure) are placed 4.88 g (20 mmoles) of phenyl (*E*)-2-phenylvinyl sulfone and 40 ml of anhydrous THF. The stirred solution is cooled to −90° with an acetone–liquid nitrogen bath and treated during 3 min with 12.0 ml of a 2.0 *M* methyllithium solution in ethyl ether (24 mmoles, containing LiBr). During a 30-min period, the reaction mixture is warmed to −60°, during which time methane evolution is completed. Then the lithium reagent is assayed by injecting 2.18 ml (35 mmoles) of methyl iodide. After 5 min the cooling bath is removed and the mixture held at 20–25° for 30 min. Hydrolysis, separation and drying of the organic layer, removal of solvent, and recrystallization of the residue from hexane give 4.0 g (79%) of phenyl (*E*)-1-phenyl-1-propen-2-yl sulfone, mp 93–95° [Eq. (43)].

Repetition of the above lithiation, but addition of cyclohexanone instead of methyl iodide, at −60° gave upon workup a 60:40 mixture of the starting sulfone and the desired adduct [Eq. (42)].

However, the lithiation can be repeated, and then 24 ml of a 1 *M* solution of magnesium bromide in benzene–ethyl ether[78] is added at −70°. After 30 min at −45° the mixture is treated with 2.5 ml of cyclohexanone and allowed to warm to 20–25°. Addition of aqueous ammonium chloride and ether, followed by the usual workup, gives a crude product whose nmr spectrum should show the absence of starting material. Recrystallization of the product from an ether–hexane pair gives 5.2 g (76%) of 1-[(*E*)-2-phenyl-1-phenylsulfonyl-1-ethenyl]cyclohexanol [Eq. (42)], mp. 122–124°.

[78] The magnesium bromide solution is prepared by charging a 250-ml three-necked flask (Fig. 19, with drain) with 2.6 g (0.10 g-atom) of magnesium turnings, 50 ml of dry ethyl ether, and 25 ml of dry benzene. Then 18.8 g (0.10 mole) of ethylene bromide is added dropwise. The mixture is heated at reflux for 2 hr and then diluted with 100 ml of ether. Aliquots can be analyzed by the Volhard method.

D. Compounds of Group IIB

1. General Preparative and Analytical Methods

The organometallic compounds of zinc, cadmium, and mercury have an easily discernible gradation in chemical stability. While many organozinc compounds inflame in air, decompose exothermically with protic solvents, and distill without decomposition, those of mercury are generally stable to oxygen and hydroxylic solvents and show great thermal lability. Organocadmium compounds take an intermediate position between these extremes. Both zinc and mercury compounds played an extraordinary role in the development of organometallic chemistry. Much of the chemistry we now associate with magnesium and lithium reagents was originally worked out by Saytzeff and his successors with zinc compounds.[79] In addition, the atmospheric stability of organomercury compounds destined them for early experimental study, and we find that as early as 1853[80] one of the founders of organometallic chemistry, Edward Frankland, devised syntheses for these compounds.

The limited reactivity of these metal alkyls and the high toxicity of mercury compounds have curtailed their use. However, in the research laboratory certain derivatives will continue to be of special value, for example (*a*) zinc alkyls for converting tertiary carbon halides into quaternary carbon centers,[81] and halomethylzinc halides as methylenating agents[82]; (*b*) organocadmium halides as reagents for converting acyl halides into ketones[83]; and (*c*) halomethylmercury compounds as methylenating agents,[84] and diorganomercury compounds for preparing more reactive, solvate-free organometallics[85]:

$$\tfrac{n}{2}R_2Hg + M \longrightarrow R_nM + \tfrac{n}{2}Hg \qquad (44)$$

---

[79] E. Krause and A. von Grosse, "Die Chemie der Metallorganischen Verbindungen," pp. 14–68. Gebrüder Borntraeger, Berlin, Germany, 1937.

[80] E. Frankland, *Justus Liebigs Ann. Chem.* **85,** 361 (1853).

[81] C. R. Noller, *J. Am. Chem. Soc.* **51,** 594 (1929).

[82] G. Wittig and K. Schwarzenbach, *Justus Liebigs Ann. Chem.* **650,** 1 (1961).

[83] J. Cason, *Chem. Rev.* **40,** 15 (1947).

[84] D. Seyferth, J. M. Burlitch, R. J. Minasz, J. Y.-P. Mui, H. D. Simmons, A. J.-H. Treiber, and S. R. Dowd, *J. Am. Chem. Soc.* **87,** 4259 (1965).

[85] E. Frankland and D. F. Duppa, *Justus Liebigs Ann. Chem.* **130,** 118 (1864).

The most frequently employed preparative methods[86] for Group IIB organometallics include (*a*) the interaction of the metal with an organic halide or with an organoidal halide, such as chloromethylsilane [Eq. (45)]; (*b*) a metal–metal exchange between the metal salt and a more reactive organometallic [Eq. (46)]; (*c*) a proton–mercury exchange with carbon acids and mercuric salts [mercuration, Eq. (47)]; (*d*) a metal–mercury displacement with a R_2Hg and a more active metal [Eq. (48)]; and (*e*) a redistribution reaction of an organometallic with its metal salt [Eq. (49)]:

$$\mathrm{EMX} \xleftarrow{\mathrm{EX}} \mathrm{M} \xrightarrow{\mathrm{RX}} \mathrm{RMX} \tag{45}$$

$$\mathrm{MX_2} \xrightarrow{\mathrm{RM'}} \mathrm{RMX} \xrightarrow{\mathrm{RM'}} \mathrm{R_2M} \tag{46}$$

$$\mathrm{R{-}C{\equiv}C{-}H} \xrightarrow{\mathrm{HgZ_2}} \mathrm{(R{-}C{\equiv}C{-})_2Hg} \tag{47}$$

$$\mathrm{M + R_2Hg \longrightarrow R_2M + Hg} \tag{48}$$

$$\mathrm{R_2M + MX_2 \longrightarrow 2RMX} \tag{49}$$

There is no generally applicable qualitative test for the presence or constitution of these individual organometallics. But these unassociated, covalent compounds can be subjected to the chromatographic and spectrometric analyses applicable to ordinary organic compounds if due consideration is taken of their thermal and atmospheric instability.

Quantitative determination of the metal content generally involves complete hydrolytic conversion to the inorganic salt and then a customary titrimetric or gravimetric determination of the metal content, although direct iodimetric determination of carbon–zinc bonding has been developed.[87]

2. Ethylzinc Halide

Organozinc halides can be prepared, in certain cases, directly from the organic halide and zinc dust or from a 1:1 interaction of zinc halide and the appropriate Grignard reagent. However, if the zinc alkyl or aryl is available, the cleanest and most convenient method is the redistribution reaction with the anhydrous zinc halide[88]:

$$\mathrm{R_2Zn + ZnX_2 \longrightarrow 2RZnX} \tag{50}$$

This method avoids the presence of any undesired metal alkyl or its salts and allows a vaıiety of solvents to be chosen.

---

[86] K. Nützel, *in* "Methoden der Organischen Chemie" (E. Müller, ed.), Vol. XIII/2a, Organometallic Compounds, pp. 553–858, 859–950, Georg Thieme Verlag, Stuttgart, Germany, 1973.

[87] N. Barychnikov and A. A. Kvasov, *Z. Anal. Chem.* **249,** 212 (1970).

[88] A. Job and R. Reich, *Bull. Soc. Chim. Fr.* **33,** 1414 (1923).

Procedure[89]:

$$(CH_3CH_2)_2Zn + ZnX_2 \longrightarrow CH_3CH_2ZnX \quad (51)$$

In a Schlenk flask equipped with a pressure-equalized funnel (Fig. 13c) anhydrous zinc chloride prepared as described in Section D,3, or anhydrous zinc bromide prepared from ethylene bromide and zinc dust, is treated with two equivalents of diethylzinc. The mixture is heated at 70° until the zinc halide has dissolved (15–30 min). Then the excess diethylzinc is distilled off under reduced pressure. The residue can be dissolved in dry, deoxygenated ethyl ether, THF, benzene, or a nonhydroxylic solvent. Alternatively, the ethylzinc halide can be recrystallized from dry, deoxygenated pentane to give over 90% of the pure product: chloride, mp 66–68°; bromide, mp 80–81°; or iodide, mp 96–99° (dec).

The nmr spectra of the ethylzinc halides in toluene show the expected methylene quartet and methyl triplet, with the methylene quartet centered at (δ) 0.65 (Cl), 0.73 (Br), and 0.80 (I) ppm. With the iodide the value had to be obtained graphically because of disproportionation.

3. Diethylzinc

The preparation of diethylzinc is intimately associated with the very origins of organometallic chemistry, for Frankland[90] first prepared this compound in the course of his search for free alkyl radicals. In the prototypical organometallic synthesis, ethylzinc iodide was formed from the alkyl iodide and zinc metal. Then, by thermal disproportionation, the zinc dialkyl itself was produced:

$$2CH_3CH_2I + 2Zn \longrightarrow 2CH_3CH_2ZnI \xrightarrow{\Delta} (CH_3CH_2)_2Zn + ZnI_2 \quad (52)$$

Not only in their modes of formation but also in their reactions zinc alkyls have set the pattern for the application of metal alkyls in organic synthesis.

The ready commercial accessibility and the relatively low cost of certain aluminum alkyls recommend them as convenient transalkylating agents for preparing zinc alkyls.[91]

Procedure:

$$Zn + 2HCl \xrightarrow{-H_2} ZnCl_2 \xrightarrow{2(CH_3CH_2)_3Al} (CH_3CH_2)_2Zn + [(CH_3CH_2)_2AlCl]_2 \quad (53)$$

The anhydrous zinc chloride is prepared in a 500-ml Schlenk vessel (Fig. 13c) surmounted by a reflux condenser, which via a calcium chloride drying

---

[89] (a) J. Boersma and J. G. Noltes, *Tetrahedron Lett.* 1521 (1966); (b) J. Boersma and J. G. Noltes, *J. Organometal. Chem.* **8,** 551 (1967).

[90] E. Frankland, *Justus Liebigs Ann. Chem.* **71,** 171 (1849); **85,** 347 (1853); **95,** 28 (1855).

[91] J. J. Eisch, *J. Am. Chem. Soc.* **84,** 3605 (1962).

tube is connected to an acid trap. The side neck is provided with a gas inlet tube that dips into the liquid. The reaction vessel is charged with 32.7 g (0.50 g-atom) of pure zinc shavings and 125 ml of anhydrous ether. With magnetic stirring dry hydrogen chloride gas is led into the suspension continuously until no more metal is visible (3–4 hr). The viscous, colorless biphasic mixture is warmed *in vacuo* to remove the ether and dissolved gas. Finally, the resulting white mass is cautiously melted by heating the chloride with a luminous flame or hot-air blower.

The side neck of the Schlenk vessel is provided with a nitrogen inlet, and the main neck with a pressure-equilized addition funnel containing 150 ml (1.10 moles) of redistilled triethylaluminum. (The whole apparatus is carefully flushed beforehand with nitrogen.) Then the aluminum reagent is added slowly to the zinc chloride. Although there is slight gas evolution, little heat is generated. After occasional shaking the mixture becomes almost clear, with some suspended gray particles (zinc metal from hydride in the aluminum reagent).

The addition funnel is removed, and the flask attached to a distillation apparatus as shown in Fig. 29, except that the single Schlenk vessel used as a receiver is cooled in an acetone–solid carbon dioxide bath. At a pressure of 15–20 mm Hg the diethylzinc is distilled over at 23–25° and solidifies in the receiver, 62 g (78%).

The nmr spectrum of neat diethylzinc shows its methylene quartet at (δ) 0.30 and its methyl triplet at 1.15 ppm.[92]

4. Di-1-alkynylmercury

Terminal acetylenes can be readily converted to diorganomercury compound by Nessler's solution, an alkaline solution of the tetraiodomercurate(II) ion. These air- and moisture-stable compounds have sharp melting points and serve as useful derivatives for isolating and identifying these hydrocarbons.[93] Subsequent exchange reactions with aluminum hydrides or more reactive organometallics can be utilized for the preparation of other alkynylmetallics[94]:

$$(C_6H_5{-}C{\equiv}C{-})_2Hg + 2R_2AlH \longrightarrow 2C_6H_5{-}C{\equiv}C{-}AlR_2 + Hg\downarrow + H_2 \qquad (54)$$

Procedure:

$$2R{-}C{\equiv}C{-}H + K_2HgI_4 + 2NaOH \longrightarrow (R{-}C{\equiv}C{-})_2Hg + 2KI + 2NaI + 2H_2O \qquad R = CH_3, C_6H_5 \qquad (55)$$

---

[92] D. F. Evans and J. P. Maher, *J. Chem. Soc.* 5125 (1962).

[93] (a) M. Kutscheroff, *Ber.* **17,** 13 (1884); (b) J. R. Johnson and W. L. McEwen, *J. Am. Chem. Soc.* **48,** 469 (1926).

[94] J. J. Eisch and M. W. Foxton, unpublished studies (1967).

Caution: Although not very volatile, these toxic organomercury compounds should be handled and stored in closed systems, and disposable gloves should be worn.

The reagent, prepared from 66 g (0.24 mole) of mercuric chloride and 163 g (1.0 mole) of potassium iodide in 163 ml of water and made alkaline with 125 ml of 10% aqueous sodium hydroxide solution, is stirred and cooled in a 500-ml two-necked flask to 0–5°, while propyne gas is passed into the mixture, until a precipitate no longer forms. The solid is collected and washed with 50% aqueous ethanol (the filtrate can be retreated with propyne to ensure complete reaction) to give 54 g (80%) of di-1-propynylmercury which is recrystallized from methanol, mp 202–204°.

A liquid alkyne, such as phenylacetylene, is first redistilled, 0.20 mole is dissolved in 200 ml of 95% ethanol, and the solution added dropwise to the foregoing mercuric iodide solution. The same workup gives a comparable yield of bis(phenylethynyl)mercury, which recrystallizes from 95% ethanol as glistening leaflets, mp 125–126°.

5. Bis(trimethylsilyl)mercury

This compound exemplifies the class of organoidal metallic compounds, since it has carbonlike groups, $(CH_3)_3Si$ [cf. $(CH_3)_3C$], bonded to a metal. Such compounds can be prepared by (*a*) the reaction of the metal with organoid halides or ethers (e.g., R_3SiCl or R_3SiOR' and Li); (*b*) the cleavage of metal–metal bonds by metals (e.g., $R_3Si{-}SiR_3$ and M); and (*c*) the interconversion of one organoidal metallic into another by treatment with a metal salt (e.g., $R_3SiLi + R'_3SnCl \rightarrow R_3Si{-}SnR'_3 + LiCl$).[95] The following illustration of the first method produces a most useful reducing agent for various organic and organometallic halogen derivatives.[96,97]

Procedure:

$$2(CH_3)_3SiCl + 2Hg(Na) \longrightarrow (CH_3)_3Si{-}Hg{-}Si(CH_3)_3 + 2NaCl \qquad (56)$$

Caution: Mercury and its organic derivatives are extremely toxic. Bis(trimethylsilyl)mercury is not only somewhat volatile, but air readily oxidizes it to mercury. All operations should be carried out in an efficient hood, disposable gloves should be worn, and all glassware cleaned with hot, concentrated nitric acid promptly after use.

All operations and storage of product should be done under an atmosphere of pure, dry nitrogen. The amalgamation process is exothermic

---

[95] (a) D. Wittenberg and H. Gilman, *Q. Rev.* **13,** 116 (1959); (b) H. Gilman and G. L. Schwebke, *Adv. Organometal. Chem.* **1,** Chapter 3 (1964).

[96] G. Neumann and W. P. Neumann, *J. Organometal. Chem.* **42,** 277 (1972).

[97] J. J. Eisch and H. P. Becker, *J. Organometal. Chem.* **171,** 141 (1979).

and should be conducted with good stirring and gradual admixing. Before use, the reaction vessels should be inspected carefully for any scratches or weak spots. Stirring and shaking operations should take place in deep metal trays whose volume can retain the contents in the event of breakage.

In a 500-ml three-necked flask (Fig. 19, without drain) equipped with a motor-driven blade stirrer, a reflux condenser, and an addition funnel (whose drop rate can be finely adjusted) are placed 6.0 g (0.26 g-atom) of freshly sliced sodium and 30 ml of dry toluene. The suspension is stirred under reflux to produce finely dispersed sodium. The heating is stopped, and then 60 ml (860 g, 4.3 g-atoms) of pure mercury is very slowly added from the funnel, so as to keep the stirred toluene suspension at moderate reflux. (*Caution: Strong initial exotherm.*) After all the mercury has been added, the stirred suspension is cooled, the stirrer assembly removed, the toluene pipetted off, and the residual toluene removed *in vacuo*.

A thick-walled hydrogenation bottle (150 ml) whose mouth has an inner ground joint is provided with a two-necked adapter, one neck stoppered and the other bearing a nitrogen connection. After flushing with nitrogen the bottle is charged with the liquid amalgam (by pipet) and with 28.4 g (0.26 mole) of freshly distilled chloro(trimethyl)silane. (*Caution: Easily hydrolyzed and irritating.*) The adapter is then quickly removed, and the bottle sealed with a well-greased stopper secured in place by springs or rubber bands. The bottle is shaken mechanically for 3 days, during which time dark-yellow particles of product are formed. The bottle is then opened, and the two-necked adapter quickly reattached. *Caution: Any yellow product adhering to the stopper will usually ignite as it is removed. Therefore no flammable solvents should be in the vicinity.* (Alternatively, the following operations can be done in a dry box.)

The product is extracted in portions into 300 ml of dry, oxygen-free benzene (until extracts are only slightly yellow), and the extracts filtered through a glass frit of medium porosity (the Schlenk vessel in Fig. 13d equipped with a filter). The transfer of the benzene and its extracts is best done by a large nitrogen-flushed syringe. The filtered extracts are freed of solvent *in vacuo* to yield 30 g (67%) of yellow bis(trimethylsilyl)mercury, mp 100–102°. The air sensitivity of the product serves as a basis for analysis: A stock solution can be prepared by redissolving the solid in dry, oxygen-free benzene, and the concentration determined by simply exposing 1.0-ml aliquots to the air in tared 5-ml beakers and weighing the mercury after washing with ether and drying.

6. Phenylmercuric Bromide

Alkylmercuric halides can be prepared by a wide variety of methods, among which are (*a*) the reaction of mercury or sodium amalgam with an organic halide,[98] (*b*) the cleavage of R_2Hg with halogen or hydrogen halide, (*c*) the redistribution reaction between R_2Hg and HgX_2,[99] (*d*) the oxymercuration of olefins with mercuric acetate and conversion of the adduct into the mercuric halide with X^-,[100] and (*e*) the monoalkylation of HgX_2 with Grignard or organolithium compounds.[101] The last method is the most versatile for arylmercuric halides as well, although two methods unique to aryl derivatives are the decomposition of aryldiazonium mercuric salts, $ArN{=}N^+\ HgCl_3^-$,[102] and the electrophilic mercuration of the aromatic nucleus.[103] An important transmetalation is that between arylboronic acids and mercuric chloride.[104] The resulting arylmercuric chloride is a useful sharp-melting derivative for characterizing the organic group, and the amount of hydrogen chloride set free in water can serve as a quantitative analytical method for the purity or neutralization equivalent of the boronic acid.[105]

Procedure[106]*:*

$$C_6H_5MgBr + HgBr_2 \longrightarrow C_6H_5HgBr + MgBr_2 \qquad (57)$$

Caution: All mercury salts are extremely toxic. Gloves should be worn during all manipulations. Although not volatile, these finely divided mercury compounds can form solid aerosols. Transfers should be done in an efficient hood.

In a 1-liter, three-necked flask equipped as in Fig. 19 are placed 21.9 g (0.9 g-atom) of magnesium turnings and 150 ml of dry ether. After flushing with nitrogen a 20-ml portion of a solution of 133.5 g (0.85 mole) of bromobenzene in 350 ml of ether is added. After turbidity and an exotherm signal the initiation, the balance of the bromobenzene is added so as to maintain a gentle reflux. (A loose cloth collar around the flask and above the level of the ether can be packed with ice and thus supplement the condenser.) At the close of the reaction the solution is drained off into a similarly equipped flask previously flushed with nitrogen (95% yield by titration).

---

[98] J. Maynard, *J. Am. Chem. Soc.* **54,** 2108 (1932).

[99] E. Krause and A. von Grosse, "Die Chemie der Metallorganischen Verbindungen," p. 129, Verlag Borntraeger, Berlin, Germany, 1937.

[100] J. Chatt, *Chem. Rev.* **48,** 7 (1951).

[101] P. Pfeiffer and P. Truskier, *Ber.* **37,** 1125 (1904).

[102] A. N. Nesmeyanov, *Ber.* **62,** 1010, 1018 (1929).

[103] O. Dimroth, *Ber.* **31,** 2154 (1898).

[104] A. Michaelis and P. Becker, *Ber.* **15,** 182 (1882).

[105] See G. Wittig, G. Keicher, A. Rückert, and P. Raff, *Justus Liebigs Ann. Chem.* **563,** 110 (1949) for the titrimetric procedure for both qualitative and quantitative analyses of arylboranes by this so-called sublimate method.

[106] S. Hilpert and G. Grüttner, *Ber.* **46,** 1675 (1913).

The mercuric bromide is ground finely and then dried overnight *in vacuo*. With vigorous stirring, 360 g (1.0 mole) of the bromide is added in small portions to the Grignard solution (Fig. 23). The resulting suspension is stirred under reflux for 8 hr, the organic layer drained off (negative Gilman color test I) and the residue boiled with three 250-ml portions of 1% aqueous hydrogen chloride solution (to remove residual $MgBr_2$ and $HgBr_2$). The crude product is washed successively with hot water, ethanol, and ether, each time in a thorough manner, to yield 286 g (94%) of a fluffy solid. After prolonged drying *in vacuo* at 100° the phenylmercuric bromide melts at >270°. This product can be recrystallized from pyridine to yield compact platelets, mp 280°, but the *thoroughly dried* unrecrystallized product is better suited for heterogeneous reactions, such as the preparation of phenylboron dibromide from boron tribromide.

7. Diphenylmercury

The three most useful methods for preparing diorganomercury compounds are (*a*) the treatment of mercuric salts with organomagnesium,[101,107] -lithium,[108] -aluminum,[109] or -boron[110] compounds; (*b*) the reaction of organic halides with sodium amalgam[111]; and (*c*) the reduction of organomercuric halides.[112] The low solubility of mercuric halides in ether at 25° retards the reaction with Grignard reagents. For this reason, the use of a THF medium has proved advantageous.[113]

Procedure:

$$2C_6H_5MgBr + HgBr_2 \xrightarrow{THF} (C_6H_5)_2Hg + MgBr_2 \tag{58}$$

Caution: Diphenylmercury is extremely toxic. Although it is not as volatile as mercury alkyls, it does have a perceptible odor. All transfers should be carried out in a closed system or in an efficient hood. Disposable gloves should be worn, and all glassware brought into contact with mercury compounds should be cleaned promptly after use with hot, concentrated nitric acid.

In a 2-liter three-necked apparatus equipped as in Fig. 19 are placed 75.8 g (3.13 g-atoms) of magnesium turnings and 100 ml of dry, freshly

[107] W. E. Bachmann, *J. Am. Chem. Soc.* **55,** 2830 (1933).

[108] H. Gilman and R. E. Brown, *J. Am. Chem. Soc.* **52,** 3314 (1930).

[109] H. Jenkner, Kali-Chemie, F.R. Germany Patent 1,048,481 (1956) [*Chem. Abstr.* **54,** 24401 (1960)].

[110] J. B. Honeycutt and J. M. Riddle, *J. Am. Chem. Soc.* **81,** 2593 (1959).

[111] F. Fuchs, *J. Prakt. Chem.* **119,** 209 (1928).

[112] H. Gilman and M. M. Barnett, *Rec. Trav. Chim.* **55,** 563 (1936).

[113] J. J. Eisch and W. C. Kaska, *J. Am. Chem. Soc.* **88,** 2976 (1966).

distilled THF. After initiation, a solution of 162 g (1.04 moles) of bromobenzene in 450 ml of THF is added over a 3-hr period. The solution is stirred under reflux, cooled, and drawn off through the filter into a 3-liter flask preflushed with nitrogen (as in Fig. 19, but without filter drain). A 450-ml portion of dry benzene is added, and 150 g (0.416 mole) of solid mercuric bromide is introduced in portions into the vigorously stirred Grignard reagent over a 2-hr period (Fig. 23). The mixture is stirred at reflux for 120 hr and then slowly treated with 300 ml of a saturated aqueous ammonium chloride solution. (An inert atmosphere is not needed after this point.) The separated aqueous solution is extracted with two 100-ml portions of benzene, the organic extracts combined and dried over anhydrous calcium sulfate, and the solvent evaporated to yield 125 g (85%) of colorless needles, mp 123–125°.[114]

Although this product is suitable for most purposes, absolute assurance that any phenylmercuric bromide is absent can be gained in the following manner. The mercury product is added to a solution of 25 ml of hydrazine hydrate (*caution: extremely toxic*) in 650 ml of absolute ethanol, and the mixture heated at reflux for 7 hr. The ethanol is distilled off, and the solvent traces removed *in vacuo*. Dissolution of the solid residue and filtration remove the finely divided mercury. Concentration of the filtrate and cooling yield colorless needles, mp 124.5–125.5° (70% overall yield). Decomposition of a sample in hot nitric acid and addition of aqueous silver nitrate solution give no sign of halogen.

The foregoing procedure has been applied, on the same scale, to the preparation of di-*m*-tolylmercury; the yield of halogen-free product, mp 101–102°, is 110 g (69%).

8. Bis[(*E*)-stilbenyl]mercury and Bis[(*Z*)-stilbenyl]mercury

The generation of vinylic lithium reagents from vinylic bromides with retention of configuration can best be achieved by halogen–lithium exchange with lithium alkyls at low temperatures.[115] These stereochemically defined lithium reagents can react in turn with various metallic halides to yield organometallics of known configuration. The procedure illustrates the individual synthesis of E and Z isomers of the stilbenylmercury system.[116]

Of the known vinylic lithium reagents, those bearing aryl groups are

---

[114] V. S. Petrosyan and O. A. Reutov [*Zh. Org. Khim.* **3**, 2074 (1967)] have studied the effect of solvent on the ortho-proton–^{199}Hg coupling and the chemical shift of the meta and para protons relative to the ortho proton: 7.2 ppm in cyclohexane ranging to 7.9 ppm in hexamethylphosphorus triamide (HMPT).

[115] G. Köbrich, *Angew. Chem.* **79,** 15 (1967).

[116] A. N. Nesmeyanov, A. E. Borisov, and N. A. Vol'kenau, *Izv. Nauk SSR Otdel. Khim. Nauk* 992 (1954) [*Chem. Abstr.* **49,** 6892 (1955)].

the most prone to geometric isomerization. (*Z*)-Stilbenyllithium is known to isomerize to the E isomer at higher temperatures or in the presence of stronger donors, such as THF. Thus the treatment of (*E*)-α-bromostilbene with *n*-butyllithium at $-110°$ and subsequent carbonation give a 58% yield of (*E*)-α-phenylcinnamic acid. When the reaction with *n*-butyllithium is done at $-10°$, a 61% yield of (*Z*)-α-phenylcinnamic acid is obtained[117a]:

$$\underset{\text{E isomer}}{(C_6H_5)(H)C{=}C(C_6H_5)(COOH)} \xleftarrow[\text{2. } H^+]{\text{1. } CO_2} \underset{\text{Z isomer}}{(C_6H_5)(H)C{=}C(C_6H_5)(Li)} \xrightarrow{-10°} (C_6H_5)(H)C{=}C(Li)(C_6H_5) \xrightarrow[\text{2. } H^+]{\text{1. } CO_2} \underset{\text{Z isomer}}{(C_6H_5)(H)C{=}C(COOH)(C_6H_5)} \quad (59)$$

Procedure:

$$(C_6H_5)(H)C{=}C(C_6H_5)(Br) \xrightarrow[-n\text{-BuBr, } -40°]{n\text{-BuLi}} (C_6H_5)(H)C{=}C(C_6H_5)(Li) \xrightarrow[-\text{LiCl}]{\frac{1}{2}HgCl_2} [(C_6H_5)(H)C{=}C(C_6H_5)]_2Hg \quad (60)$$

In a 200-ml Schlenk flask equipped with a pressure-equalized addition funnel and a nitrogen inlet are placed 5.2 g (20 mmoles) of (*E*)-α-bromostilbene,[117b] 20 ml of dry benzene, and 40 ml of anhydrous ethyl ether. The magnetically stirred mixture is cooled to $-40°$ and then treated dropwise with 14 ml of 1.6 *M* *n*-butyllithium in hexane. After 60 min at $-40°$, 3.0 g (10 mmoles) of dried, powdered mercuric chloride is added in one portion. After 30 min of vigorous stirring the mixture is treated with 50 ml of 5% aqueous hydrogen chloride. Separation and drying of the organic layer over anhydrous sodium sulfate give, upon removal of the solvent, a crude product that is recrystallized from a benzene–95% ethanol pair. Bis[(*E*)-stilbenyl]-mercury is obtained in 55% yield (3.1 g), mp 145–147°.

Procedure:

$$(C_6H_5)(H)C{=}C(Br)(C_6H_5) \xrightarrow[-n\text{-BuBr}]{n\text{-BuLi}} (C_6H_5)(H)C{=}C(Li)(C_6H_5) \xrightarrow[-\text{LiCl}]{\frac{1}{2}HgCl_2} [(C_6H_5)(H)C{=}C(C_6H_5)]_2Hg \quad (61)$$

On the same scale as the foregoing reaction, (*Z*)-α-bromostilbene[118] is converted into bis[(*Z*)-stilbenyl]mercury in 70% yield, mp 243–245° (from benzene–ethanol).

---

[117] (a) D. Y. Curtin and W. J. Koehl, *J. Am. Chem. Soc.* **84,** 1967 (1962); (b) This compound is prepared by the addition of bromine to *trans*-stilbene and the monodehydrobromination of this adduct with alcoholic potassium hydroxide: W. M. Jones and R. Damico, *J. Am. Chem. Soc.* **85,** 2273 (1963).

[118] This compound is obtained by the addition of hydrogen bromide to diphenylacetylene in acetic acid solution at 0°: G. Drefahl and K. Ponsold, *Chem. Ber.* **93,** 505 (1960).

E. Compounds of Group IIIA

1. General Preparative and Analytical Methods

The reactivity of Group IIIA organometallic compounds diverges from the pattern shown by the compounds of Groups IA and IIA, where reactivity increases with the metal's atomic weight. In Group IIIA air-oxidizability generally decreases with atomic weight, with aluminum compounds, R_3M, being converted readily into $M(OR)_3$ and those of boron, gallium, indium, and thallium into R_2MOR. In protodemetalation boron compounds also deviate: Although reaction with water occurs with decreasing ease in going from aluminum to thallium, many boron alkyls are unaffected by deoxygenated water (other types, such as boron aryls or alkynyls, however, can be hydrolyzed). Finally, the Lewis acidity of these metal alkyls shows a similar pattern: As exemplified by coordination with ethyl ether, such acidity decreases from aluminum to thallium. In not ordinarily forming etherates, boron compounds are anomalous. The exceptional behavior of organoboron compounds in all these comparisons is chiefly attributed to steric hindrance in a reagent's access to the carbon–boron bond.[119a]

Despite these wide variations in reactivity, the compounds of boron,[119b] aluminum,[120] and thallium[121] have each been shown to possess a rich organic chemistry of their own. Their individual virtues in organic synthesis stem from key preparative paths to the crucial intermediate: (*a*) for boron, the addition of the metal hydride to carbon–carbon unsaturation [Eq. (62)]; (*b*) for aluminum, the addition of the metal alkyl to carbon–carbon unsaturation [Eq. (63)]; and (*c*) for thallium, the hydrogen–metal (metalation) exchange reaction [Eq. (64)]:

$$\gt M{-}H + {-}C{\equiv}C{-} \longrightarrow (H)\gt C{\cdots}C\lt(M) \tag{62}$$

$$\gt M{-}R + {-}C{\equiv}C{-} \longrightarrow (R)\gt C{\cdots}C\lt(M) \tag{63}$$

$$\gt M{-}Z + H{-}C\lessgtr \longrightarrow \gt M{-}C\lessgtr + H{-}Z \tag{64}$$

---

[119] (a) H. C. Brown, *J. Chem. Educ.* **36,** 424 (1959); (b) "Organic Syntheses via Boranes," Wiley (Interscience), New York, 1975.

[120] T. Mole and E. A. Jeffrey, "Organoaluminium Compounds," Elsevier, Amsterdam, 1972.

[121] A. McKillop and E. C. Taylor, *Adv. Organometal. Chem.* **11,** 147 (1973).

As important as these reactions are in organic synthesis, they are not unique to the individual metals but may be realized with compounds of other Group IIIA metals as well. The hydrometalation reaction [Eq. (62)] has also been observed with hydrides of aluminum, gallium, and indium.

Other important preparative methods for these compounds include[122] (*d*) metal–metal exchanges between Group IIIA halides and other metal alkyls, R_nM', where the reactivity of R_nM' can vary greatly [from RLi to R_4Sn or RHgX, Eq. (65)]; (*e*) metal–metal displacement between the metal (Al, Ga, In) and R_2Hg [Eq. (66)]; (*f*) redistribution between the metal alkyl and its salt [Eq. (67)]; (*g*) thermal rearrangement or cyclization reactions [Eq. (68)]; and (*h*) the preparation of mixed derivatives of R_3M by partial oxidation, reduction, or solvolysis [Eq. (69)]:

$$nMX_3 + 3R_nM' \longrightarrow nR_3M + 3M'X_n \tag{65}$$

$$2M + 3R_2Hg \longrightarrow 2R_3M + 3Hg \tag{66}$$

$$2R_3M + MX_3 \longrightarrow 3R_2MX \tag{67}$$

$$\text{2-biphenylyl–Al(Ph)}_2 \xrightarrow[-\text{PhH}]{\Delta} \text{9-phenyl-9-aluminafluorene (Al–Ph)} \tag{68}$$

$$R_2MOR \xleftarrow{\frac{1}{2}O_2} R_3M \xrightarrow[-RH]{R'OH} R_2MOR' \tag{69}$$

With such great variation in reactivity, one cannot expect any broadly applicable methods for direct assay of the carbon–metal content of these compounds. However, their covalent character does permit the use of chromatographic and spectral analysis if their general air and moisture sensitivity and the light and thermal lability of indium and thallium compounds are taken into account. Complicating such analysis is also the tendency of aluminum compounds to exist as dimers or trimers and the general associated character of mixed alkyls of the type R_nMZ_{3-n}.

Aside from the various methods involving the complete hydrolysis of R_3M to an inorganic salt and its subsequent analysis,[123] organoboranes of the most diverse type can be quantitatively oxidized and estimated with trimethylamine oxide,[124] and organoalanes (R_3Al and R_2AlH) can be completely hydrolyzed and the resulting hydrocarbon measured by volume,

---

[122] See E. Müller (ed.), "Methoden der Organischen Chemie," Vol. XIII/4, Organometallic Compounds, Georg Thieme Verlag, Stuttgart, Germany, 1970.

[123] See T. R. Crompton, "Chemical Analysis of Organometallic Compounds," Vol. 1–5, Academic Press, New York, 1973–1977.

[124] R. Köster and Y. Morita, *Justus Liebigs Ann. Chem.* **704,** 70 (1967).

mass spectrometry, or gas chromatography.[125] Finally, attention should be drawn to the specialized methods for the analysis of boron hydride[126] and alkylaluminum hydride solutions.[125,127]

2. 1-Methyl-4,5-dihydroborepin and 1-Chloro-4,5-dihydroborepin

As in the synthesis of *pentaphenylborole* (Section E,8), synthesis of the dihydroborepin system ensues readily when the corresponding tin heterocycle is admixed with the appropriate boron halide. Thus the use of phenylboron dichloride or phenylboron dibromide readily yields 1-phenyl-4,5-dihydroborepin, but the distillative separation from the by-product, dihalo(di-*n*-butyl)tin, is difficult.[128] However, the use of boron trichloride or methylboron dibromide, without solvent, surmounts this problem, and essentially quantitative yields result.[129]

Procedure:

$$c\text{-}C_6H_8B\text{—}Cl \xleftarrow[-(n\text{-}C_4H_9)_2SnCl_2]{BCl_3} c\text{-}C_6H_8Sn(n\text{-}C_4H_9)_2 \xrightarrow[-(n\text{-}C_4H_9)_2SnBr_2]{CH_3BBr_2} c\text{-}C_6H_8B\text{—}CH_3 \quad (70)$$

Caution: Methylboron derivatives are highly volatile and pyrophoric. With a positive nitrogen pressure in a methylboron-containing vessel, the release of an aperture or removal of a stopper will cause the emerging vapor to ignite (with a green flame). Gastight syringes should be used for transfers, and all methylboron-containing vessels should be well chilled before opening.

In the C chamber of a dual-chamber vessel as shown in Fig. 24 (chamber capacity of 50 ml; other neck of C provided with a septum; other neck of A provided with a nitrogen inlet) are placed 23.7 g (75.7 mmoles) of 1,1-di-*n*-butyl-4,5-dihydrostannepin. Chamber C is cooled in an acetone–solid carbon dioxide bath to $-78°$, and the stirred stannepin rapidly treated with 7.05 ml (75.2 mmoles) of methylboron dibromide, which is injected through the septum. After the vigorous reaction has subsided, chamber C is brought to room temperature and chamber A cooled to $-78°$. Chamber C is warmed

---

[125] H. Lehmkuhl and K. Ziegler, *in* "Methoden der Organischen Chemie" (E. Müller, ed.), Vol. XIII/4, Organometallic Compounds, pp. 284–309, Georg Thieme Verlag, Stuttgart, Germany, 1970.

[126] Reference 119b, pp. 239–251.

[127] Reference 125, pp. 293–297.

[128] D. Sheehan, Doctoral Dissertation, Yale Univ., University Microfilms, Ann Arbor, Michigan (1964).

[129] J. J. Eisch, J. E. Galle, and R. J. Wilcsek, unpublished results (1974).

to 95–100° for 20 min and then cooled to 25°. The pressure in the apparatus is reduced to 0.1 mm Hg, and chamber C warmed to transfer the 1-methyl-4,5-dihydroborepin to A, 7.0 g (88%), which is pure as shown by nmr spectroscopy. Upon cooling C, the dibromo(di-*n*-butyl)tin solidifies.

The nmr spectrum of 1-methyl-4,5-dihydroborepin shows absorptions at (δ, CCl_4) 1.03 (s, 3H), 2.55 (broad t, 4H, $J = 2.5$ Hz), 6.39 (d, 2H, $J =$ 13 Hz) and 7.02 (broad d, 2 H, $J = 13$ Hz) ppm.

In a similar manner, 1-chloro-4,5-dihydroborepin can be prepared by condensing 9.2 g (78 mmoles) of boron trichloride in chamber A, which is cooled to −78°. Then 22 g (70 mmoles) of 1,1-di-*n*-butyl-4,5-dihydrostannepin is injected into the similarly cooled chamber C. Allowing A gradually to warm up while still cooling C introduces the boron chloride into the stannepin. After reaction is complete, the borepin is transferred to A to give a yield of 8.5 g (96%) of 1-chloro-4,5-dihydroborepin. Chamber C contains 11 g of fine needles of dichloro(di-*n*-butyl)tin.[130]

The nmr spectrum of 1-chloro-4,5-dihydroborepin (pentane) shows absorptions at 2.38 (broad t, 4 H), 6.22 (d, 2 H, $J = 13$ Hz), and 7.10 (broad d, 2 H) ppm.

3. Phenylborane–Triethylamine

For the successful preparation of organoboron hydrides care must be taken to suppress the tendency of these mixed boranes to disproportionate:

$$3RBH_2 \longrightarrow R_3B + B_2H_6 \qquad (71)$$

This reaction can be avoided by working at low temperatures and complexing the boron hydride with a tertiary amine. For example, although the reduction of phenylboron dichloride with lithium aluminum hydride in refluxing THF yielded only diborane and triphenylborane,[131] a similar reduction of diethyl phenylboronate (phenylboron diethoxide) at −70° in the presence of pyridine gave a 78% yield of phenylborane–pyridine.[132,133]

Organoboron hydrides are most suitable for preparing boracyclic compounds from various dienes, especially if five-, six-, or seven-membered heterocycles can be formed. Either the amine complexes of organoboron

---

[130] Sheehan conducted this synthesis in benzene but experienced great difficulty with frothing upon attempted separation from solvent. The reaction can be conducted either without solvent, as above, or with the reagents separately dissolved in pure, dry pentane. Separation can then be effected by drawing off the borepin and pentane under reduced pressure and fractionating the volatile fraction to isolate the dihydroborepin [J. J. Eisch and A. E. Radkowsky, unpublished studies (1970)].

[131] D. R. Nielson, W. E. McEwen, and C. A. Van der Werf, *Chem. Ind.* **37,** 1069 (1957).

[132] M. F. Hawthorne, *J. Am. Chem. Soc.* **80,** 4291 (1958).

[133] N. N. Greenwood and J. C. Wright, *J. Chem. Soc.* 448 (1965).

dihydrides or dimeric dialkylboron hydrides have been employed; in the latter instance, intermediate polymers or the diadduct disproportinates thermally[134,135]:

$$\text{R}\sim\text{(borolane, B–R')}\sim\text{R} \xleftarrow[\Delta]{R'BH_2\cdot NR''_3} \text{R–CH=CH–CH=CH–R} \xrightarrow{(R'_2BH)_2} R'_2B\text{–(CH}_2)_4\text{–}BR'_2 \xrightarrow[-R'_3B]{\Delta} \text{(borolane, B–R')} \quad (72)$$

Procedure[132,133]:

$$C_6H_5BCl_2 \xrightarrow[-2HCl]{2CH_3CH_2OH} C_6H_5B(OCH_2CH_3)_2 \xrightarrow[(CH_3CH_2)_3N]{LiAlH_4,\ Et_2O} C_6H_5BH_2\cdot N(CH_2CH_3)_3 \quad (73)$$

Diethyl phenylboronate can be prepared[136,137] in essentially quantitative yield in the following manner: In a 250-ml Schlenk flask (Fig. 13c, under nitrogen, with a pressure-equalized addition funnel) are placed 50 g (0.31 mole) of phenylboron dichloride and then, with cooling to 0°, 40 g (0.87 mole) of absolute ethanol (dried over anhydrous $MgSO_4$) is added very slowly with stirring. (Provision should be made to trap the large amount of HCl evolved.) After stirring overnight, the mixture is distilled through a 10-cm Vigreux column to yield 56 g of product, bp 55–56° at 0.7 mm. The nmr spectrum shows absorptions at (CCl_4) 1.22 (t, 6 H), 4.05 (q, 4 H), 7.27 (m, 3 H), and 7.55 (m, 2 H) ppm.

In a 500-ml three-necked flask (Fig. 19, without drain or condenser) are placed 1.65 g (44 mmoles) of powdered lithium aluminum hydride, 300 ml of anhydrous ethyl ether, and 17 g (0.17 mole) of triethylamine (freshly distilled from pellet KOH). To this mixture, which is stirred at −70°, is added dropwise a solution of 10.8 g (60 mmoles) of diethyl phenylboronate in 50 ml of ether. The mixture is allowed to warm to 0° over a period of 12 hr. Then the mixture is cautiously poured onto 300 ml of ice water. The ether layer is decanted through a filter, the aqueous layer extracted with ether, and the combined and dried ($MgSO_4$) organic extracts evaporated down to a volume of 30–40 ml. Cooling to −78° deposits colorless needles, mp

---

[134] (a) R. Köster, G. Griasnow, W. Larbig, and P. Binger, *Justus Liebigs Ann. Chem.* **672,** 1 (1964); (b) R. Köster, "Progress in Boron Chemistry," Vol. I, pp. 289–344. Pergamon, Oxford, 1964.

[135] (a) H. C. Brown, E. Negishi, and P. L. Burke, *J. Am. Chem. Soc.* **94,** 3561 (1972); (b) H. C. Brown and E. Negishi, *J. Am. Chem. Soc.* **94,** 3567 (1972).

[136] P. B. Brindley, W. Gerrard, and M. F. Lappert, *J. Chem. Soc.* 1540 (1956).

[137] E. W. Abel, W. Gerrard, and M. F. Lappert, *J. Chem. Soc.* 3833, 5051 (1957).

63–64°, 6.9 g (60%). The infrared spectrum of the hydride shows an intense hydride band at (CCl_4) 2325 cm^{-1}.

4. Phenylboron Dibromide and Diphenylboron Bromide

Organoboron halides can be prepared by (*a*) redistribution reactions between R_3B and BX_3,[138] (*b*) reactions between boronic or borinic acid anhydrides and BX_3,[137] and (*c*) transmetalation reactions between mercury or tin alkyls and BX_3.[139,140] Either the easily prepared phenylmercuric bromide (Section C) or commercially available tetraphenyltin is useful for synthesizing phenylboron dihalides. For diphenylboron halides, the reaction of the tin aryl with BX_3 gives much higher yields than the reaction of diphenylmercury with BX_3.[141,142]

Procedure[139,142]*:*

Caution: All boron halides hydrolyze readily in moist air, hence are highly irritating. Although the arylboron halides are not sensitive to oxidation, a dry, inert atmosphere is required. Since silicone stopcock grease is readily attacked, lengthy storage of these halides should be in sealed ampuls.

$$C_6H_5HgBr + BBr_3 \xrightarrow{\Delta} C_6H_5BBr_2 + HgBr_2 \quad (74)$$

In a 2-liter three-necked flask equipped as in Fig. 19 are placed 286 g (0.8 mole) of thoroughly dried, fluffy, finely divided (unrecrystallized) phenylmercuric bromide and 1000 ml of dry benzene. Under a dry nitrogen atmosphere 200 g (0.8 mole) of boron tribromide is introduced into the vigorously stirred slurry. After a reflux period of 12 hr the cooled benzene solution is filtered and the filtrate collected in a Schlenk flask (Figs. 26 and 13c). The benzene is largely removed under reduced pressure, and the residue pipetted into a distillation apparatus (Fig. 29). After a forerun the phenylboron dibromide is collected at 63–65° at 1.0 mm Hg (the vessel should be chilled to avoid sublimation of the bromide). Redistillation of this fraction provides 154 g (78%) of pure, colorless product, mp 30–32° (lit.[143] 32–34°). After drying, the recovered mercuric bromide can be rephenylated as in Section D.

The following procedure can be adapted to produce either phenylboron dibromide or diphenylboron bromide.[141]

---

[138] J. C. Lockhart, *Chem. Rev.* **65,** 131 (1965).
[139] W. Gerrard, M. Howarth, E. F. Mooney, and D. E. Pratt, *J. Chem. Soc.* 1592 (1963).
[140] J. E. Burch, W. Gerrard, M. Howarth, and E. F. Mooney, *J. Chem. Soc.* 4916 (1960).
[141] H. Nöth and H. Vahrenkamp, *J. Organometal. Chem.* **11,** 399 (1965).
[142] A. Michaelis and E. Richter, *Justus Liebigs Ann. Chem.* **315,** 19 (1915).
[143] A. Michaelis and P. Becker, *Ber.* **13,** 58 (1880).

Procedure:

$$3C_6H_5BBr_2 + C_6H_5SnBr_3 \xleftarrow{3BBr_3} (C_6H_5)_4Sn \xrightarrow{2BBr_3} (C_6H_5)_2BBr + SnBr_4 \qquad (75)$$

In a 300-ml Schlenk vessel equipped with a pressure-equalized dropping funnel (Fig. 13c) are placed 42.7 g (101 mmoles) of dried, finely ground tetraphenyltin and 50 ml of purified, dry methylene chloride. The whole system is flushed well with dry nitrogen and cooled to −78° in an acetone–solid carbon dioxide bath. Then a solution of 50.1 g (202 mmoles) of boron tribromide in 50 ml of methylene chloride is added slowly in dropwise portions to the stirred suspension for the purpose of obtaining diphenylboron bromide. (*Caution: Extremely vigorous reaction.*)

If phenylboron dibromide is desired, then 73 g (303 mmoles) of boron tribromide in 50 ml of methylene chloride is employed.

The resulting mixture is allowed to warm up to room temperature gradually and with constant stirring over a 2-hr period. (*Caution: Further exotherm upon warming.*) The methylene chloride is distilled off, and the temperature of the residue slowly raised to 200–225°. (For the preparation of phenylboron dibromide, the residue is distilled directly.) After 2 hr, the mixture is distilled to yield a forerun of stannic bromide and a main fraction of diphenylboron bromide, bp 150–160° at 8 mm Hg, 38 g (76%).

For the preparation of phenylboron dibromide, the product distills at 99–101° at 20 mm Hg, mp 32–34°. The nmr spectrum of phenylboron dibromide shows absorptions at (δ, CCl_4) 7.2–7.65 (m, 3 H, meta and para protons) and 8.1–8.3 (d of d, 2 H, ortho protons) ppm.

5. 1-Phenylboracyclopentane

α,ω-Alkadienes and their analogs, such as divinyl ethers and divinylsilanes, can be converted into boron heterocycles by various boron hydrides. The most convenient reagents seem to be tertiary amine complexes of phenylborane[133] or t-butylborane[144] and the dimeric diethylboron hydride.[134]

$$\text{c-}E(CH_2)_4B{-}CH_2CH_3 \xleftarrow[-(CH_3CH_2)_3B]{[(CH_3CH_2)_2BH]_2} E(CH{=}CH_2)_2 \xrightarrow[-R_3N]{R'{-}BH_2\cdot NR_3} \text{c-}E(CH_2)_4B{-}R' \qquad (76)$$

Since the products are unsymmetric organoboranes, they are prone to thermal disproportionation. To minimize this rearrangement, the temperature should be kept as low as practical during the hydroboration and purification processes.

---

[144] M. F. Hawthorne, *J. Am. Chem. Soc.* **83**, 2541 (1961).

Procedure:

$$C_6H_5BH_2 \cdot N(CH_2CH_3)_2 + CH_2{=}CH{-}CH{=}CH_2 \xrightarrow{\Delta} \underset{\substack{| \\ C_6H_5}}{\text{(cyclic } C_4H_8B)} + (CH_3CH_2)_3N \qquad (77)$$

In a 200-ml three-necked flask, equipped with a regular condenser surmounted by a Dewar condenser cooled to −78°, a gas inlet tube dipping into the reaction solution, and a nitrogen inlet, are placed 8.0 g (46 mmoles) of the phenylborane–triethylamine complex and 20 ml of dry benzene. With stirring at reflux under nitrogen, the solution is slowly treated with a stream of 6 g (0.11 mole) of gaseous butadiene, which is dispensed from a cold trap by gradual warming. After all the butadiene has been absorbed, the solution is cooled and transferred to a dual-chamber apparatus (Fig. 24, chamber A capacity of 50 ml). Then at 0° all volatiles are removed under high vacuum. Chamber A is allowed to warm up to room temperature, while chamber C is cooled to −30°. The 1-phenylboracyclopentane is slowly evaporated and condensed into C. The yield of product melting at −18 to −20° is 3.7 g (60%). A complete analysis of the neat infrared spectrum of this heterocycle has been made.[133]

6. Triphenylborane

Triarylboranes are generally prepared by the action of an aryl Grignard[145] or lithium[146] reagent on boron trifluoride etherate or a trialkyl borate $B(OR)_3$. A convenient variant for storing a source of triphenylborane is to employ an excess of the phenylmetallic and thereby form $MB(C_6H_5)_4$. Treatment of the latter in aqueous solution with trimethylamine hydrochloride precipitates $[(CH_3)_3NH]^+[B(C_6H_5)_4]^-$, an air- and water-stable salt, which can be pyrolyzed as required to yield the boron aryl.[147]

Procedure[147]*:*

$$3C_6H_5Li + BF_3 \cdot O(CH_2CH_3)_2 \longrightarrow 3LiF + (C_6H_5)_3B$$

$$\downarrow \text{1. } C_6H_5Li \quad \text{2. } (CH_3)_3NHCl$$

$$(C_6H_5)_3B + (CH_3)_3N + C_6H_6 \xleftarrow{\Delta} [(CH_3)_3NH]^+[B(C_6H_5)_4]^- \qquad (78)$$

In a 1-liter three-necked apparatus equipped as in Fig. 19, but without the filter drain, are placed 28.5 g (0.20 mole) of freshly distilled boron trifluoride etherate and 20 ml of anhydrous ether. Then 335 ml of a 1.8 *M*

[145] (a) E. Krause and H. Polack, *Ber.* **59,** 777 (1926); (b) E. Krause and R. Nitsche, *ibid.* **55,** 1261 (1922).
[146] G. Wittig, G. Keicher, A. Rückert, and P. Raff, *Justus Liebig Ann. Chem.* **563,** 110 (1949).
[147] G. Wittig and P. Raff, *Justus Liebigs Ann. Chem.* **573,** 195 (1951).

ethereal solution of phenyllithium (0.60 mole) is added dropwise with stirring and cooling. The reaction mixture is then heated at reflux for 4 hr. The resulting biphasic system is transferred in portions by syringe under nitrogen to the distillation apparatus shown in Fig. 29, except that a short-path air condenser (1.0 cm i.d.) is used. The ether is removed by water aspiration (using a moisture trap), and the residual product in the flask is gradually heated up to 160–180° in a Woods Metal bath to volatilize the biphenyl. By means of a smoky flame or heat gun, the biphenyl is driven into one of the cooled receivers. By raising the bath temperature to 250–300° (at a pressure of 15–20 mm Hg) the triphenylborane distils over as a pale-yellow, viscous liquid, bp 200–210°. Heating the air condenser prevents clogging by solidified product. The triphenylborane is obtained in 65–75% yield (33–36 g) and is sufficiently pure for most purposes, mp 138–140°. Recrystallization from ether raises the melting point to 142–142.5° (146–147°).

For the preparation of trimethylammonium tetraphenylborate, the foregoing procedure is simply modified in the following manner. Instead of 0.60 mole, one adds 0.80 mole of phenyllithium solution. After the reflux period, Gilman's color test I is found to be negative. (An inert atmosphere is not necessary after this point.) The ether is volatized away from the reaction mixture *in vacuo*, and the solid residue is extracted thoroughly with four 75-ml portions of water (biphenyl is left behind). These extracts are admixed with an aqueous solution of 0.21 mole of trimethylamine hydrochloride to precipitate the ammonium tetraphenylborate. The salt is washed with water and dried in a vacuum desiccator (90%). The *thoroughly dried* salt can be pyrolyzed in the same distillation apparatus described above. First, under nitrogen at atmospheric pressure the salt is heated at 200–210° to cause decomposition. After evolution of the amine and benzene has ceased, a stream of nitrogen is used to sweep these volatiles from the system. The triphenylborane is then distilled at 15 mm, bp 245–265°, to yield 80–90% of product, based on the starting ammonium complex.

The nmr spectrum of triphenylborane shows (δ, CCl_4) a multiplet centered at 7.4 ppm and ranging from 7.2 to 7.7 ppm.

7. Diphenyl(phenylethynyl)borane–Pyridine

Unsymmetric organoboranes can be prepared by (*a*) the reaction of organoboron halides or esters with various organolithium,[147] -magnesium,[145] or -aluminum reagents[148]; (*b*) the hydroboration of olefins or

[148] R. Köster, *Justus Liebigs Ann. Chem.* **618,** 31 (1958).

acetylenes with R_2BH[149]; and (*c*) the haloboration of olefins[150] or acetylenes[151,152] with R_2BX. In all cases, care must be taken to avoid disproportionation of the product, R_2BR', into R_3B and R'_3B. Such steps as low-temperature isolation techniques and the use of tetracoordinate complexes increase the chances of a successful synthesis. The use of aminoalkoxyborinates, $R_2B—OCH_2CH_2NH_2$, or pyridine complexes of the halide, is especially advantageous.[153,154]

Procedure[153]:

$$(C_6H_5)_2B(\leftarrow \ddot{N}C_5H_5)—Br + C_6H_5—C{\equiv}C—Li \longrightarrow (C_6H_5)_2B(\leftarrow \ddot{N}C_5H_5)—C{\equiv}C—C_6H_5 + LiBr \quad (79)$$

In a 500-ml three-necked flask equipped as in Fig. 19, but without the filter drain, are placed 110 ml of purified, anhydrous THF. With cooling to 0° and stirring, 20 g (81 mmoles) of diphenylboron bromide is added; then 7.0 ml (85 mmoles) of anhydrous pyridine (reagent grade dried by reflux over BaO) is slowly admixed, causing precipitation of the colorless pyridine complex.

In the meantime, a 200-ml Schlenk flask (Fig. 13c), equipped with a pressure-equalized addition funnel, is charged with 8.3 g of freshly distilled phenylacetylene and 50 ml of THF and the contents cooled to 0°. Then 37 ml of 2.21 *M* *n*-butyllithium in hexane are added dropwise to the acetylene solution. After 15 min the dark solution of phenylethynyllithium is transferred by syringe to the addition funnel of the first apparatus and added dropwise to the stirred boron bromide suspension. The suspension becomes a clear, dark-red solution during a further 15-min stirring period. Then 200 ml of a saturated, aqueous ammonium chloride solution is added, and the layers separated. The water layer is extracted thrice with ether, and the combined organic extracts dried over anhydrous sodium sulfate. The solvent is removed *in vacuo*, and the residual solid treated with 100 ml of methanol. The precipitated product is triturated under this solvent layer at 0° for 30 min and then filtered. After washing with methanol and drying in a vacuum desiccator,

---

[149] H. C. Brown, "Boranes in Organic Chemistry." Cornell Univ. Press, Ithaca, New York, 1972.

[150] F. Joy, M. F. Lappert, and B. Prokai, *J. Organometal. Chem.* **5,** 506 (1966).

[151] M. F. Lappert and B. Prokai, *J. Organometal. Chem.* **1,** 384 (1963).

[152] J. J. Eisch and L. Gonsior, *J. Organometal. Chem.* **8,** 53 (1967).

[153] J. Soulie and P. Cadiot, *Bull. Soc. Chim. Fr.* 1981 (1966).

[154] B. M. Mikhailov and V. A. Vaver, *Izv. Akad. Nauk SSSR* 812 (1957) [*Chem. Abstr.* **52,** 3667 (1958)].

the colorless diphenyl(phenylethynyl)borane–pyridine melts at 159–160°, 16.5 g (59%).

The nmr spectrum of this pyridine complex shows absorptions at (δ, $CDCl_3$) 7.1–7.5 (m, 17 H), 7.76 (d, 1 H, gamma-pyridyl H, $J = 7.5$ Hz) and 8.74 (d of d, 2 H, alpha-pyridyl H, $J = 7.5$ and 1.5 Hz) ppm.

8. Pentaphenylborole

This heterocycle can be synthesized by the action of phenylboron dihalide on (*E,E*)-1,2,3,4-tetraphenyl-1,3-butadien-1,4-ylidenedilithium (Section B,8), but its isolation cannot succeed in the manner originally described.[43] Contrary to the first claimed synthesis, pentaphenylborole cannot tolerate transfers in air or recrystallization from alcohol. As formed under these reaction conditions, pentaphenylborole exists as a pale-yellow etherate that dissociates upon warming to yield the deep-blue (with admixed salts, an apparent deep-green) heterocycle. Manipulated in air, the solvated borole is converted into products yielding tetraphenylfuran and 3-benzylidene-1,2-diphenylidene upon hydrolysis.[155]

Although the original procedure can be adapted to take account of the air sensitivity of pentaphenylborole, separation of lithium salts and organic by-products can be tedious. Hence the simplicity of the procedure using phenylboron dichloride and 1,1-dimethyl-2,3,4,5-tetraphenylstannole is most appealing.[155,156]

Procedure[157]*:*

$$(C_6H_5)_4C_4Sn(CH_3)_2 + C_6H_5BCl_2 \longrightarrow (C_6H_5)_4C_4B(C_6H_5) + (CH_3)_2SnCl_2 \quad (80)$$

In a 100-ml Schlenk flask (Fig. 13c, septum closure) are placed 5.04 g (10 mmoles) of 1,1-dimethyl-2,3,4,5-tetraphenylstannole and 30 ml of dry, deoxygenated toluene. Then the stirred suspension is treated by syringe with 1.33 ml (10 mmoles) of phenylboron dichloride. The mixture changes rapidly from yellow to dark green and then slowly to blue-green, at which stage (~1.5 hr) it should be filtered. The toluene is siphoned off by an immersion filter (Fig. 14, a siphon fused to a coarse filter). The blue solid is washed thrice with 10-ml portions of toluene and dried *in vacuo* to yield 3.2 g (72%)

---

[155] J. J. Eisch, N. K. Hota, and S. Kozima, *J. Am. Chem. Soc.* **91,** 4575 (1969).

[156] J. J. Eisch and J. E. Galle, unpublished studies (1974).

[157] (a) This is an improved procedure developed by Dr. J. E. Galle (1974); (b) See G. E. Herberich, J. Hengesbach, U. Kölle, and W. Oschmann, *Angew. Chem. Int. Ed.* **16,** 42 (1977).

of deep-blue pentaphenylborole. The solid is sensitive to oxygen, moisture, and stopcock grease, and hence should be used within a week of its preparation; under nitrogen it melts over 240° with decomposition.

9. Heptaphenyl-7-borabicyclo[2.2.1]heptadiene

The 1,3-diene character of metalloles permits 7-metallabicyclo[2.2.1]-heptadienes to be isolated in certain cases. That this borole undergoes a rapid Diels–Alder reaction at 25° with the relatively unreactive dienophile, diphenylacetylene, points up its high dienic reactivity.[158] The silicon analog reacts readily with such dienophiles as benzyne, dimethyl acetylenedicarboxylate, and phenylacetylene,[44] but not with diphenylacetylene. The analogous germole[159] and stannole[160] derivatives react with phenylacetylene at high temperatures to produce pentaphenylbenzene. The bisphosphine complexes of 2,3,4,5-tetraphenylnickelole are in fact intermediates in the catalytic trimerization of diphenylacetylene by nickel(0).[161]

Procedure[158]*:*

$$\text{pentaphenylborole } (C_6H_5)_4C_4B{-}C_6H_5 + C_6H_5{-}C{\equiv}C{-}C_6H_5 \longrightarrow \text{heptaphenyl-7-borabicyclo[2.2.1]heptadiene} \quad (81)$$

Into a 200-ml Schlenk flask (Fig. 13c, septum closure) containing 2.16 g (4.87 mmoles) of pentaphenylborole is injected a solution of 890 mg (5.0 mmoles, free of *trans*-stilbene) of diphenylacetylene in 25 ml of dry, deoxygenated toluene. The suspension is stirred at 20–25° until the dark-blue color has disappeared. The mixture is diluted with 75 ml of pure, dry ether to precipitate the colorless heptaphenyl-7-borabicyclo[2.2.1]heptadiene, mp 210° (dec) (60%).

The nmr spectrum of this bicyclic borane shows (δ, $CDCl_3$) a multiplet from 7.2 to 7.8 ppm, which has a strong singlet at 7.51 ppm. The ultraviolet spectrum in ethyl ether has its longest wavelength maximum at 318 nm ($\varepsilon = 10^5$).

[158] J. J. Eisch and J. E. Galle, *J. Am. Chem. Soc.* **97,** 4436 (1975).
[159] N. K. Hota and C. J. Willis, *J. Organometal. Chem.* **15,** 89 (1968).
[160] K. Kuno, K. Kobayashi, M. Kawanisi, S. Kozima, and T. Hitomi, *J. Organometal. Chem.* **137,** 349 (1977).
[161] J. J. Eisch and J. E. Galle, *J. Organometal. Chem.* **96,** C23 (1975).

10. Heptaphenylborepin

Diels–Alder adducts of metalloles and acetylenes show a strong tendency to undergo thermal aromatization with elimination of the metal bridge.[44,159,160] Such reactions are reductive eliminations, for the metal unit formed has had its valence diminished by 2. Indeed, the pyrolysis of 7-silabicyclo-[2.2.1]heptadienes has served as a way to generate silicon(II) or silene intermediates.[44] With such 7-borabicyclic derivatives, two modes of reaction have been observed: (*a*) Tetracoordinate borate anions undergo bridge rupture and rearrange to arylborate anions[162]; and (*b*) the tricoordinate boron system undergoes sequential suprafacial, sigmatropic, and disrotatory ring-opening rearrangements[163] to yield the borepin.[158]

Procedure[158]:

$$C_6H_5B(C_6H_5)_6C_7 \xrightarrow{\Delta} (C_6H_5)_7\text{-borepin} \quad (82)$$

In a 100-ml Schlenk vessel equipped with a reflux condenser, 1.39 g (2.23 mmoles) of heptaphenyl-7-borabicyclo[2.2.1]heptadiene is heated at reflux in 30 ml of dry, deoxygenated toluene for 24 hr, during which period the initially colorless mixture turns a brilliant fluorescent green. Most of the toluene is removed *in vacuo*, and the residue is diluted with dry ether to yield 1.18 g (84%) of bright-green heptaphenylborepin, mp 233–236°.

The nmr spectrum of this borepin shows (δ, $CDCl_3$) absorptions from 6.2 to 7.5 ppm, with an intense singlet at 7.0 ppm. The electronic spectrum in ethyl ether shows maxima (ε) at 412 (6100), 342 (8080), 276 (22,700), and 245 (28,000) nm.

11. Salts of the Bis(2,2′-biphenylene)borate Anion

As an example of the isolation of countless known tetraorganoborate salts, the synthesis and interconversion of various cationic salts of this spiroborate anion have several interesting aspects. The requisite 2,2′-dilithiobiphenyl[164] need no longer be made from the "2,2′-biphenylene-

---

[162] J. J. Eisch and J. E. Galle, *J. Organometal. Chem.* **127,** C9 (1976).

[163] R. B. Woodward and R. Hoffmann, "The Conservation of Orbital Symmetry," Verlag Chemie-Academic Press, Weinheim, Germany, 1971.

[164] G. Wittig and W. Herwig, *Chem. Ber.* **88,** 962 (1955).

mercury" of Wittig (actually a tetramer)[165] but is readily accessible from 2,2′-dibromobiphenyl via a lithium–bromine exchange with *n*-butyllithium. The lithium tetraarylborate can in turn be converted into the sodium, ammonium, tetramethylammonium, or potassium salt by suitable precipitations.[166] In fact, the air- and water-stable sodium tetraphenylborate itself is a highly prized analytical reagent for the gravimetric determination of the heavier alkali metal, ammonium, or other cations.[167]

The spiro structure of the bis(2,2′-biphenylene)borate anion permits suitably nuclear-substituted derivatives to be resolved into enantiomeric anions[168] and also offers an interesting test system for studying rearrangements of tetraorganoborates.[169]

Procedure[169]*:*

2 (2,2′-dibromobiphenyl; Br Br) $\xrightarrow[-2n\text{-}C_4H_9Br]{2n\text{-}C_4H_9Li}$

2 (2,2′-dilithiobiphenyl; Li Li) $\xrightarrow[-3LiF]{BF_3\cdot OEt_2}$ Li^+ [bis(2,2′-biphenylene)borate, B]$^-$ (83)

In a 500 ml three-necked flask equipped as in Fig. 19 are placed 23.4 g (75 mmoles) of 2,2′-dibromobiphenyl and 150 ml of anhydrous ethyl ether. The contents are stirred at 0° while 57 ml of a 2.8 *M* solution of *n*-butyllithium in hexane (160 mmoles) is added dropwise. The mixture is then stirred at room temperature for 5 hr. The resulting solution is transferred to a nitrogen-flushed, pressure-equalized addition funnel connected to a 1-liter three-necked flask equipped as in Fig. 19 (without drain) and containing 5.1 g (36 mmoles) of freshly distilled boron trifluoride etherate and 50 ml of ether. The lithium reagent is added dropwise to the stirred boron trifluoride at room temperature (slight exotherm), and the mixture then heated at reflux for 20 hr. A semisolid, white precipitate now coats the wall of the

[165] G. Wittig and W. Herwig, *Chem. Ber.* **87,** 1511 (1954).
[166] G. Wittig and G. Lehmann, *Chem. Ber.* **90,** 875 (1957).
[167] W. Gerrard, "The Organic Chemistry of Boron," Academic Press, New York, 1961.
[168] K. Torssell, *Acta Chem. Scand.* **16,** 87 (1962).
[169] (a) J. J. Eisch and R. J. Wilcsek, *J. Organometal. Chem.* **71,** C21 (1974); (b) J. J. Eisch and R. J. Wilcsek, unpublished studies (1973).

flask. The whole mixture is cooled to $-78°$, the ether layer is decanted through a filter, and the residue is combined with three 30-ml portions of ether and washed repeatedly at $-78°$. All solids are combined and then shaken with 70 ml of acetone. The suspension is filtered through a pad of Celite on a coarse glass frit to remove the insoluble lithium fluoride. Concentration of the filtrate deposits more lithium fluoride. The filtrate can be transferred to the dual-chamber apparatus in Fig. 24 and by the usual vapor technique the lithium bis(2,2′-biphenylene) borate can be recrystallized, first from acetone and then from ether. During recrystallization a temperature of $-10°$ must be maintained, lest the product form an oil. After drying *in vacuo*, a 45–50% yield of colorless solid is obtained, whose analysis shows the presence of about one molar equivalent of admixed lithium fluoride in the product.

To prepare the ammonium salt, the original filtered and concentrated acetone solution is added dropwise to a stirred solution of 15 g of ammonium chloride in 200 ml of water. The suspension is stirred for 30 min and filtered. The collected solid is washed with 30 ml of water and then dissolved in 20 ml of acetone. After filtration, the filtrate is diluted with 200 ml of water and some yellow, semisolid precipitate removed. Concentration to a volume of 150 ml and cooling deposit a colorless, granular solid. Further concentration gives a second crop of ammonium bis(2,2′-biphenylene)borate, which after drying *in vacuo* over phosphorus pentoxide amounts to 5–6 g (45%).

To prepare the sodium salt, 5.0 g (15 mmoles) of this ammonium salt is added, in one portion, to a solution of 15 mmoles of sodium methoxide in 100 ml of anhydrous methanol contained in a 250-ml Schlenk flask. After 1 hr at room temperature the solution is heated at reflux for 3 hr, while a stream of nitrogen from the gas inlet adapter sweeps the evolved ammonia up the condenser and out of the system. The methanol is evaporated away, and the residue treated with 100 ml of purified, dry 1,2-dimethoxyethane. After heating to reflux, the solution is filtered and cooled to yield nicely crystalline sodium bis(2,2′-biphenylene)borate, which after drying at 35° *in vacuo* has a melting point of 123–135° (dec). Elemental and nmr spectral analyses show this product to be a tris(1,2-dimethoxyethane)solvate, 6.0 g (67%): In acetone-d_6 nmr absorptions are displayed at (δ) 3.27 (s, 18H, CH_3), 3.45 (s, 12H, CH_2), 6.7–7.1 (m, 12H), and 7.67 (d of t, 4H, $J = 7.0$ and 1.4 Hz) ppm. For analysis, the product is recrystallized twice from the same solvent.[170]

To prepare the potassium salt, the original pasty, crude lithium salt (before the acetone treatment) is extracted with four 100-ml portions of water

[170] J. J. Eisch and K. Tamao, unpublished studies (1975).

and the extracts filtered. Also, the decanted ether solution is extracted with two 50-ml portions of water. The combined aqueous extracts are treated with 4.5 g (60 mmoles) of potassium chloride. The solution is concentrated down to ~150 ml to precipitate a granular solid. The collected solid is dissolved in a minimum of acetone, and the remaining inorganic salts filtered off. The acetone is evaporated, and the residue triturated at 0° with two 75-ml portions of ether. The remaining solid, upon drying at 65° *in vacuo* amounts to 8 g (60%) of potassium bis(2,2'-biphenylene)borate. From a chloroform–methanol pair this salt crystallizes as a mass of fluffy needles.

To prepare the tetramethylammonium salt (or sodium salt), an aqueous solution of the potassium salt is treated with an aqueous solution of tetramethylammonium chloride (or NaCl). The flocculent precipitate can be recrystallized from methanol to yield glistening platelets of tetramethylammonium bis(2,2'-biphenylene)borate, mp > 360° (dec). Its nmr spectrum shows the expected ratio of aryl to methyl protons.

12. Diisobutylaluminum Deuteride

Dialkylaluminum hydrides can be prepared by (*a*) the synthesis of aluminum hydride from the elements in R_3Al solution and the redistribution of the groups attached to aluminum,[171] (*b*) the hydrogenolysis of R_3Al under pressure,[171,172] (*c*) the thermal elimination of olefin from R_3Al,[173] and (*d*) the interaction of dialkylaluminum halide with lithium hydride.[174,175] The former two methods are more suitable for industrial practice, since they require autoclaves, pressure equipment, and a large-scale operation. The latter two methods are useful in smaller-scale research applications. In fact, pure diisobutylaluminum hydride can be obtained from commercially purchased triisobutylaluminum by heating it between 120° and 180° for several hours at ordinary pressures until isobutylene gas evolution has ceased (a periodic reduction in pressure being used to remove dissolved isobutylene) [Part I, Eq. (12)]. But when the R_3Al type does not lose alkene readily or when R_2AlD derivatives are desired, then the fourth method is the most useful.

---

[171] K. Ziegler, H. G. Gellert, H. Lehmkuhl, W. Pfohl, and K. Zosel, *Justus Liebigs Ann. Chem.* **629,** 1 (1960).

[172] R. Köster, G. Bruno, and H. Lehmkuhl, U.S. Patent 3,097,066 (1960) [*Chem. Abstr.* **55,** 15350 (1961)].

[173] K. Ziegler, H. Martin, and F. Krupp, *Justus Liebigs Ann. Chem.* **629,** 14 (1960).

[174] K. Ziegler, H. G. Gellert, H. Martin, K. Nagel, and J. Schneider, *Justus Liebigs Ann. Chem.* **589,** 91 (1954).

[175] J. J. Eisch and S. G. Rhee, *J. Am. Chem. Soc.* **96,** 7276 (1974).

Procedure[175]:

$$3\left[\begin{matrix}(CH_3\underset{\displaystyle CH_3}{\underset{|}{C}}HCH_2)_2AlCl\end{matrix}\right]_2 + 6LiD \xrightarrow{Et_2O} 2\left[\begin{matrix}(CH_3\underset{\displaystyle CH_3}{\underset{|}{C}}HCH_2)_2AlD\end{matrix}\right]_3 + 6LiCl \qquad (84)$$

In a 1-liter three-necked flask equipped as shown in Fig. 19 are placed 300 ml of ethyl ether (reagent finally distilled from $LiAlH_4$) and 8.6 g (1.07 moles) of lithium deuteride. After cooling with a 0° bath, the stirred suspension is treated dropwise with 146 ml (0.75 mole) of freshly distilled diisobutylaluminum chloride. The reaction mixture is heated at reflux for 48 hr and then a 0.5-ml aliquot of the clear supernatant solution is withdrawn and hydrolyzed with dilute nitric acid solution. If the solution still gives a positive chloride test with a silver nitrate solution, then 0.5–1.0 g more of lithium deuteride is added and heating at reflux continued (24–48 hr) until the chloride test is negative. The cooled suspension is filtered under nitrogen through a sintered-glass frit of medium porosity. Most of the ether is then removed from the filtrate under reduced pressure, and the turbid liquid refiltered. The resulting diisobutylaluminum deuteride is heated in an oil bath at 75° and at a reduced pressure of 2 mm Hg to remove any coordinated ether. After 4 hr the residue is distilled through a Vigreux column to yield ~110 ml of the deuteride (bp 110–125° at 0.5 mm Hg), whose nmr spectrum shows the presence of 8% ethyl ether. A careful refractionation yields ~90 ml (67%) of ether-free diisobutylaluminum deuteride, bp 112–118° at 0.5 mm Hg. By use of the apparatus depicted in Fig. 10, 3.0 ml of absolute ethanol is placed in the flask, cooled with a liquid-nitrogen trap, and treated with 0.50 ml of the distilled deuteride. The system is evacuated, and then the cold bath removed to permit solvolysis. Mass spectral analysis of the collected gas shows the presence of 96% deuterium hydride. The hydride content can also be determined by modifications[175] of the Bonitz isoquinoline titration,[176] which permits R_2AlH to be measured in the presence of R_3Al and R_2AlOR.[125,127]

The infrared spectrum shows a broad, intense absorption at (cm^{-1}, neat) 1175–1350 (Al—D), and only a weak residual band is detectable at 1740 (Al—H).

13. Diisobutyl[(*Z*)-1-trimethylsilyl-2-phenylethenyl]aluminum

Unsymmetric vinylic aluminum compounds can be prepared by (*a*) the reaction of a vinylic lithium,[177] sodium,[178] or Grignard[179] reagent with an

---

[176] (a) E. Bonitz, *Chem. Ber.* **88,** 742 (1955); (b) D. F. Hagen and W. D. Leslie, *Anal. Chem.* **35,** 814 (1963); (c) J. H. Mitchen, *ibid.* **33,** 1331 (1961).

[177] J. M. Riddle, Ethyl Corp., U. S. Patent 3,262,959 (1964) [*Chem. Abstr.* **65,** 13758 (1966)].

[178] B. Bartocha, A. J. Bilbo, D. E. Bublitz, and M. Y. Gray, *Z. Naturforsch.* (*b*) **16,** 357 (1961).

[179] H. E. Ramsden, M & T Corp., U. S. Patent 3,109,851 (1959) [*Chem. Abstr.* **60,** 3015 (1964)].

organoaluminum halide; (*b*) the exchange reaction between $AlCl_3$ and vinylzinc or -lead compounds to yield vinylaluminum chlorides,[180] (*c*) the addition of a triorganoaluminum to an acetylene[181–183] and (*d*) the hydralumination of an acetylene with a diorganoaluminum hydride.[181,184,185] In these reactions, the stability of the resulting geometric isomer becomes a point of concern. Both kinetic and thermodynamic control of product configuration has been observed. The hydralumination of trimethyl(phenylethynyl)silane described here, which is typical of many facile additions possible with heteroatom-substituted acetylenes, proceeds regiospecifically to give the thermodynamically controlled trans adduct. This product has been shown to arise via cis-hydralumination and isomerization.

Procedure[185,186]*:*

$$6C_6H_5{-}C{\equiv}C{-}Si(CH_3)_3 + 2[(iso\text{-}C_4H_9)_2Al{-}H]_3 \longrightarrow 3\left[\begin{matrix} C_6H_5 \\ H \end{matrix}\!\!>\!C{=}C\!<\!\!\begin{matrix} Al(iso\text{-}C_4H_9)_2 \\ Si(CH_3)_3 \end{matrix}\right]_2 \qquad (85)$$

The diisobutylaluminum hydride should be freshly distilled and found to have a >96% hydride content as determined by the isoquinoline titration. The preparation of trimethyl(phenylethynyl)silane is described in Section F,3.

In a 200-ml two-necked flask equipped as in Fig. 13c is placed 24.5 g (0.141 mole) of the silane dissolved in 35 ml of dry pentane. While being stirred in a bath at 20–25°, the solution is treated dropwise with 20.0 g (0.141 mole) of the hydride dissolved in 25 ml of dry pentane. After stirring 8–15 hr, a sample is withdrawn, with a gastight syringe, and injected into a cooled solution of 0.5 *N* hydrochloric acid (10% EtOH). By gas chromatography, less than 1% of the starting silane and trimethyl(β-phenylethyl)silane should be present. The typical ratio of the *cis*- and *trans*-trimethyl(β-styryl)-silanes should be 2:98.

The nmr spectral absorptions characteristic of the Z-isomeric aluminum compound are (δ, neat) 0.17 (d, 4H, $J = 6.5$ Hz), 0.18 (s, 9H), 0.88 (d, 12H, $J = 6.5$ Hz), 1.70 (m, 2H), 7.23 (s, 5H), and 7.84 (s, 1H) ppm.

---

[180] B. Bartocha, U. S. Patent 3,100,217 (1959) [*Chem. Abstr.* **60,** 551 (1964)].
[181] G. Wilke and H. Müller, *Justus Liebigs Ann. Chem.* **629,** 222 (1960).
[182] J. J. Eisch and C. K. Hordis, *J. Am. Chem. Soc.* **93,** 2974, 4496 (1971).
[183] J. J. Eisch and R. Amtmann, *J. Org. Chem.* **37,** 3410 (1972).
[184] J. J. Eisch and W. C. Kaska, *J. Am. Chem. Soc.* **88,** 2213 (1966).
[185] J. J. Eisch and M. W. Foxton, *J. Org. Chem.* **36,** 3520 (1971).
[186] J. J. Eisch and S. G. Rhee, *J. Am. Chem. Soc.* **97,** 4673 (1975).

14. Diphenylaluminum Chloride and Phenylaluminum Dichloride

The preparation of organoaluminum halides, as well as alkoxides, can most simply be achieved by the redistribution reaction between triorgano-aluminum and aluminum halides or alkoxides.[187] Reactions of aluminum metal with organic halides to yield organoaluminum sesquihalides also provide commercially feasible routes to organoaluminum dihalides and diorganoaluminum halides.[187,188] This preparation of diphenylaluminum chloride is illustrative of how the ratio of reagents admixed determines the principal product in these redistributions.

Procedure[187,189]*:*

$$2[(C_6H_5)_3Al]_2 + Al_2Cl_6 \longrightarrow 3[(C_6H_5)_2AlCl]_2 \quad (86)$$

In a dry box the C chamber (100-ml capacity) of the dual-chamber apparatus shown in Fig. 24 is charged with 2.32 g (17.4 mmoles) of freshly sublimed, anhydrous aluminum chloride and 9.9 g (34.8 mmoles) of tri-phenylaluminum. Its joint to the A chamber is capped. The mixture is heated in a bath at 185–190° for 6 hr. After cooling chamber A containing dry benzene is attached to C via a sintered-glass frit of medium porosity. Benzene is poured into C, the mixture warmed to dissolve the product, and the solution filtered into A. Chamber C and the frit are removed, the solution in A con-centrated *in vacuo*, and a fresh chamber C′ containing heptane then attached. Heptane is distilled into the warm solution in A until turbidity appears. Cooling A deposits colorless needles. Recrystallization is sometimes ineffec-tive if the whole mass of solid forms too quickly. With patience, a melting point of 147–150° can be obtained. The yield is 9.8 g (86%); when a run is conducted on a fivefold scale, 43% of recrystallized product is obtained.

As an alternative procedure, the reagents can be heated in a small Schlenk flask (Fig. 13c), the flask connected to a Claisen head–air condenser adapter, and the product distilled into Schlenk vessels attached to a fraction collector, bp 150° at 0.15 mm Hg (Fig. 29). For many purposes, this distilled product is adequately pure. The colorless product turns brown in the light. By employing a 2:1 ratio of aluminum chloride and triphenylaluminum and a similar procedure, phenylaluminum dichloride can be prepared, bp 190° at 0.5 mm Hg, mp 94–95°, from benzene.

---

[187] A. V. Grosse and J. M. Mavity, *J. Org. Chem.* **5,** 106 (1940).
[188] K. Ziegler and H. Martin, *Makromol. Chem.* **18–19,** 186 (1956).
[189] J. J. Eisch and W. C. Kaska, *J. Am. Chem. Soc.* **88,** 2976 (1966).

15. Triphenylaluminum

Triarylaluminum can be prepared by (*a*) the displacement reaction between diarylmercury and aluminum metal[190]; (*b*) the reaction between aryllithium or aryl Grignard reagent and aluminum chloride,[191] with subsequent thermal decomposition of the triarylaluminum etherate; (*c*) the reaction between phenylsodium or $NaAlPh_4$ and dimethylaluminum chloride and the distillative removal of the disproportionation product, trimethylaluminum[192,193]; (*d*) the exchange reaction between triphenylboron and triethylaluminum and distillative removal of triethylboron[194]; and (*e*) the dehalogenation of arylaluminum chlorides with sodium or with either trimethylaluminum or triethylaluminum, whereby in the latter case the resulting dialkylaluminum chloride is distilled off.[195,196] For laboratory purposes, the first method recommends itself by employing air-stable reagents, avoiding donor solvents, and giving the highest melting product.

Procedure[190c,197]*:*

$$3(C_6H_5)_2Hg + 2Al \longrightarrow [(C_6H_5)_3Al]_2 + 3Hg \qquad (87)$$

Either a modified Dreikugel Apparatus provided with ground-glass joints or the dual-chamber apparatus shown in Fig. 24 can be conveniently employed. In the latter case, the two chambers each have a capacity of 500 ml. After flushing the apparatus and degassing the solvent, chamber C is charged with 61.6 g (0.174 mole) of diphenylmercury (Section D,7) (*caution: extremely toxic*), 21.8 g (0.80 g-atom) of aluminum chips (machined from ingots of 99.992% purity), and 300 ml of anhydrous xylene; chamber A is provided with 10 g of aluminum chips. Chamber C is inserted in an oil bath, and chamber A is positioned above it to serve as a reflux air condenser. After the xylene solution in C is heated at reflux for 4 days, the hot solution is decanted into A (*note:* retain the mercury pool in C), and reflux continued for 2 days. The system is cooled and, under nitrogen, flask C is replaced by a clean flask C′. The contents of A are warmed to dissolve any solid tri-

---

[190] (a) H. Gilman and K. E. Marple, *Rec. Trav. Chim.* **55,** 135 (1936); (b) A. W. Laubengayer, K. Wade, and G. Lengnick, *Inorg. Chem.* **1,** 632 (1962); (c) J. J. Eisch and W. C. Kaska, *J. Am. Chem. Soc.* **88,** 2976 (1966).

[191] G. Wittig and D. Wittenberg, *Justus Liebigs Ann. Chem.* **606,** 13 (1957).

[192] National Distillers and Chem. Co., J. F. Nobis, U. S. Patent 2,960,516 [*Chem. Abstr.* **55,** 9346 (1961)].

[193] K. Ziegler, H. Lehmkuhl, and R. Schäfer, F.R. Germany Patent 1,161,895 [*Chem. Abstr.* **60,** 12050c (1964)].

[194] R. Köster and G. Bruno, *Justus Liebigs Ann. Chem.* **629,** 89 (1960).

[195] R. Streck, *J. Organometal. Chem.* **71,** 181 (1974) and references cited therein.

[196] D. Wittenberg, *Justus Liebigs Ann. Chem.* **654,** 23 (1962).

[197] J. J. Eisch and C. K. Hordis, *J. Am. Chem. Soc.* **93,** 4496 (1971).

phenylaluminum, and the warm solution decanted into C′. Chilling C′ deposits the colorless product, 18–25 g (60–85%), mp > 230°, which gives a satisfactory aluminum analysis (calcd., 10.4; found, 10.2%). By decanting the mother liquor from C′ into A and replacing A with a fresh chamber A′ containing dry, degassed toluene, the product in C′ can be recrystallized. Toluene can be distilled into C′, recrystallization carried out, and the mother liquor decanted into A′. The process can be repeated as desired. The triphenylaluminum thereby obtained, after vacuum evaporation, forms small, colorless needles, mp 241–243°.

The nmr spectrum of triphenylaluminum shows absorptions at (δ, benzene-d_6) 8.11 (q, 2H, ortho H) and 7.4 (t, 3H, meta and para H) ppm.

The amount of heating necessary for complete reaction depends upon the purity of the reagents and the state of subdivision of the metal. Removal of a sample of the xylene solution and hydrolysis permits a test for mercury remaining in the solution (Section E,27).

16. Diphenyl(phenylethynyl)aluminum

Unsymmetric alkynylaluminum derivatives can be prepared by (*a*) the reaction of sodium acetylides with dialkyl- or diarylaluminum halides[199–202]; (*b*) the reaction of terminal acetylenes with aluminum hydrides in donor solvents[203,204] and (*c*) the reaction of terminal acetylenes with triorganoaluminums, especially with trimethyl- or triphenylaluminum.[198,205] Where applicable, the last method gives an unsolvated product under mild, simple conditions.

Procedure[198]:

$$[(C_6H_5)_3Al]_2 + 2HC{\equiv}C{-}C_6H_5 \longrightarrow [(C_6H_5)_2Al{-}C{\equiv}C{-}Ph]_2 + 2C_6H_6 \qquad (88)$$

A 250-ml two-necked flask equipped as in Fig. 13c and provided with magnetic stirring is charged with 7.4 g (29 mmoles) of triphenylaluminum in the dry box, and then 100 ml of dry benzene (from sodium) is distilled into the flask. Then 3.2 g (31 mmoles) of freshly distilled phenylacetylene is introduced into the mixture with a syringe. The yellow solution is stirred for 8 hr at 40–50°. Under a nitrogen flow, the condenser is replaced with a

---

[198] J. J. Eisch and W. C. Kaska, *J. Organometal. Chem.* **2,** 184 (1964).
[199] Ethyl Corp., E. C. Ashby *et al.*, U. S. Patent 3,020,298 (1957) [*Chem. Abstr.* **56,** 15546 (1962)].
[200] G. Wilke and H. Müller, *Justus Liebigs Ann. Chem.* **629,** 222 (1960).
[201] T. Mole and J. R. Surtees, *Aust. J. Chem.* **17,** 1229 (1964).
[202] H. Demarne and P. Cadiot, *Bull. Soc. Chim. Fr.* 216 (1968).
[203] P. Binger, *Angew Chem.* **75,** 918 (1963).
[204] J. R. Surtees, *Aust. J. Chem.* **18,** 14 (1965).
[205] T. Mole and J. R. Surtees, *Chem. and Ind.* 1727 (1963).

glass frit connected to a nitrogen-filled two-necked flask. The solution is filtered into the receiving flask, and the filtrate freed of benzene *in vacuo* for 1 hr. The receiver containing the alkynylaluminum is made chamber C of the apparatus in Fig. 24. Chamber A is flushed beforehand with nitrogen and provided with a 3:2 (v/v) mixture of dry benzene and cyclopentane. As with triphenylaluminum, the product in C is recrystallized twice to give pure diphenyl(phenylethynyl)aluminum, mp 142–144° (turning red at 141°). The yield of the once crystallized product is 85%, and of the highly pure product, 5.2 g (63%).

The infrared spectrum shows a strong band at (mineral oil) 2110 cm^{-1}. The nmr spectrum shows absorptions at (δ, benzene-d_6) 8.37 (q, 4H, ortho H of Ph_2Al), 7.55 (t, 6H, meta and para H of Ph_2Al), and 6.92–7.3 (m, 5H) ppm.

17. Tribenzylaluminum

Benzylic aluminum compounds can be prepared by (*a*) the reaction of benzylic lithium or Grignard reagents with aluminum halides[206]; (*b*) the exchange reaction between tribenzylborane and triethylaluminum, in which the volatile triethylborane is removed[194]; and (*c*) the displacement reaction between dibenzylmercury and aluminum.[206] As usually conducted, the first method yields the product as a firm solvate with ether. The solvent-free aluminum compound cannot be obtained by heating *in vacuo* without considerable decomposition. The third method is generally suitable, but temperatures above 120° must be avoided. In refluxing xylene dibenzylmercury itself decomposes into bibenzyl and mercury.

Procedure A[206]*:*

$$3(C_6H_5CH_2)_2Hg + 2Al \longrightarrow 2(C_6H_5CH_2)_3Al + 3Hg \qquad (89)$$

Into the C chamber of the apparatus shown in Fig. 24 are placed 12.0 g (31.3 mmoles) of dibenzylmercury (mp 110–111° (*caution: extremely toxic and corrosive to the skin;* rubber or disposable gloves should be worn), 6.0 g (222 mg-atoms) of aluminum turnings (99.992%), and 100 ml of dry, degassed toluene. Chamber A is positioned upright to serve as a reflux air condenser. After 3 days at reflux the warm solution is decanted into chamber A. The solution is concentrated by distilling the solvent in C. Chamber C is removed, and a fresh chamber C′ attached. Cooling of A to crystallize the product, decanting the mother liquor into C′, and redistilling toluene from C′ back into A follow for successive recrystallizations. After two recrystallizations tribenzylaluminum is obtained, 7.5 g (80%), mp 114–116°. The

[206] J. J. Eisch and J. M. Biedermann, *J. Organometal. Chem.* **30,** 167 (1971).

nmr spectrum shows absorptions at (δ, toluene-d_8) 1.48 (s, 2H) and 6.67–7.17 (m, 5H) ppm.

Since the etherate, mp 48–50°, can be made simply by the addition of diethyl ether to the foregoing product, there is no advantage here in carrying out the Grignard reaction with aluminum chloride. However, the following procedure illustrates the technique of freeing such preparations of the magnesium halide by-product. Such an approach is useful for allylic and secondary or tertiary alkylaluminum derivatives.

Procedure B[206]:

$$3C_6H_5CH_2MgCl + AlCl_3 \longrightarrow (C_6H_5CH_2)_3Al\cdot O(CH_2CH_3)_2 + 3MgCl_2 \qquad (90)$$

By use of the apparatus in Fig. 19 benzylmagnesium chloride is prepared from 63 g (0.50 mole) of benzyl chloride in 90 ml of dry ether and 14 g (0.56 g-atom) of magnesium turnings in 150 ml of ether. The addition time is 2 hr, and the yield is 85% by hydrolysis and titration.[207]

In a 2-liter three-necked flask equipped as in Fig. 19 is placed 200 ml of dry ether and, with cooling, 18.9 g (0.142 mole) of freshly sublimed aluminum chloride is added in portions. The foregoing Grignard solution (0.425 mole) is added over 30 min to the vigorously stirred solution. After stirring overnight, 200 ml of benzene is added and most of the ether distilled off. The benzene solution is drained off through the filter into a Schlenk flask, and the filtrate freed of solvent first *in vacuo* and then at 160°. The residue is extracted with xylene and with heptane, and the extracts filtered through a glass frit of medium porosity. The filtrate is freed of solvent and recrystallized (Fig. 24) from heptane. The yield is 65%; analytically pure product is obtained after three recrystallizations (34%). The nmr spectrum of tribenzylaluminum etherate shows absorptions at (δ, benzene-d_6) 0.42 (t, 6H, J = 7.0 Hz), 1.74 (s, 6H), 3.1 (q, 4H, J = 7.0 Hz), and 6.83–7.36 (m, 15H) ppm.

18. Diethylgallium Chloride

Although previous work suggests that dialkylgallium chloride may be formed by the action of hydrogen chloride on trialkylgallium diethyl etherate or on dialkylgallium hydroxide (the hydrolysis product of triethylgallium),[208] the most convenient and general method is that involving a redistribution reaction between the appropriate alkyl and the halide.[209] With care taken for an adequate heating period of the reactants, essentially quantitative yields of the mixed organometallic can be achieved.

---

[207] H. Gilman and W. E. Catlin, *Org. Syn. Coll.* **I**, 2nd ed., 471 (1958).

[208] L. M. Dennis and W. Patnode, *J. Am. Chem. Soc.* **54**, 182 (1932).

[209] J. J. Eisch, *J. Am. Chem. Soc.* **84**, 3830 (1962).

Procedure[209]*:*

$$4(CH_3CH_2)_3Ga + Ga_2Cl_6 \xrightarrow{\Delta} 6[(CH_3CH_2)_2GaCl]_x \qquad (91)$$

As in the preparation of triethylgallium, 8.81 g of gallium(III) chloride (0.050 mole) is prepared directly in a 100-ml two-necked flask. Under nitrogen 15.7 g (0.10 mole) of triethylgallium is added dropwise from a pressure-equalized addition funnel to the solid chloride over a period of 20 min. Despite the strong exotherm accompanying the formation of an almost colorless solution, it is necessary to heat the reactants at 100° for 1 hr. (Failure to do so results in a wide boiling point range upon attempted distillation.) Subsequent distillation under reduced pressure results in a narrow fraction, bp 60–62° at 2 mm Hg, 98% yield, d_{20}^{20} 1.35. In contrast to the pyrophoric and moisture-sensitive character of diethylaluminum chloride, the gallium analog does not become warm on exposure to air, nor does it react vigorously with water.

19. Diethylgallium Ethoxide

The pronounced difference in reactivity between trialkylgallium and alkoxydialkylgallium, both toward oxygen and toward alcohols, permits two general approaches to the type R_2GaOR. Although for general needs either method may be serviceable, some possible limitations should be raised. First, air-oxidation of metal alkyls has been shown to involve hydroperoxide salt intermediates and their free-radical degradation products. Hence minor carbonyl and dialkoxymetal impurities may be expected even in a carefully controlled oxidation. Second, either air-oxidation or alcoholysis as a route to R_2GaOR demands great care in drying the reagents; failure to do so leads to R_2GaOH or $(RGaO)_n$.[209]

Procedure A—Air Oxidation[209]*:*

$$2(CH_3CH_2)_3Ga + O_2 \longrightarrow 2(CH_3CH_2)_2GaOCH_2CH_3 \qquad (92)$$

A 6.3-g (0.040 mole) sample of triethylgallium and a stirring bar are placed in a 100-ml three-necked flask whose necks are provided with a thermometer immersed in the liquid (slip ring), a gas-inlet tube extending down to the surface of the liquid, and an exit tube connected through a drying tower filled with porous phosphorus pentoxide to the water aspirator. The water aspirator is provided both with the usual safety bottle and a liquid-nitrogen-cooled trap. A drying tower at least 30 × 300 mm is filled with freshly activated molecular sieves (Linde Division, Union Carbide Corp.) and connected through a Y-tube to the nitrogen gas-inlet system of the foregoing vessel so that variable amounts of air can be admixed with the nitrogen.

The triethylgallium is stirred magnetically while the aspirator is adjusted to draw a slow current of air–nitrogen through the vessel, so that a temperature of 50–60° is attained. After no further exotherm (~2 hr) the contents can be transferred to a distillation flask and distilled at 1 mm. An almost quantitative yield of colorless diethylgallium ethoxide, bp 78–79°, n_D^{20} 1.449, is obtained. The characteristic, vanilla-like odor of the alkoxide is responsible for the sweet odor associated with traces of triethylgallium in air. Although the alkoxide is patently stable in dry air, a moist atmosphere changes it gradually into a semisolid mass.

In a manner analogous to the foregoing, isobutoxy(diisobutyl)gallium (bp 161–163° at 3 mm Hg, n_D^{20} 1.458, odor of apricots) and the nondistillable *n*-decoxy(di-*n*-decyl)gallium (n_D^{20} 1.463, odor of a higher alcohol) can be prepared.

Procedure B—Ethanolysis[209]*:*

$$(CH_3CH_2)_3Ga + CH_3CH_2OH \longrightarrow (CH_3CH_2)_2GaOCH_2CH_3 + CH_3CH_3\uparrow \qquad (93)$$

A 200-ml two-necked flask equipped with a pressure-equalized addition funnel and a copper coil reflux condenser provided with a gas-inlet tube is charged with 6.3 g (0.040 mole) of triethylgallium, 35 ml of dry pentane, and a stirring bar. Then a solution of 4.6 g (0.10 mole) of freshly prepared absolute ethanol in 10 ml of dry pentane is added dropwise to the magnetically stirred solution. After the vigorous gas evolution has subsided (with ethane egress through the nitrogen line and excess-pressure release valve), the solution is heated under reflux for 0.5 hr. The pentane is removed by reduced pressure, and the somewhat turbid residue transferred to a small distillation setup. The colorless ethoxy(diethyl)gallium distils at 86–87° under 3 mm Hg in 87% yield, n_D^{20} 1.448, d_4^{20} 1.158.

20. Triethylgallium

The feasible preparation of gallium alkyls can proceed by either of two general routes: (*a*) the interaction of gallium metal with the appropriate mercury alkyl without solvent[210,211]; or (*b*) the alkylation of a gallium halide by an alkyl of lithium,[212] magnesium,[208,213] zinc,[214] or aluminum.[215] When Grignard reagents are used for alkylation, a stable, distillable monoetherate will be the result; hence unsolvated gallium alkyls cannot be procured by this method. The use of relatively expensive lithium alkyls leads to gallium alkyls in high yields, but the lithium chloride must be filtered off.[212]

---

[210] H. Gilman and R. G. Jones, *J. Am. Chem. Soc.* **62,** 980 (1940).

[211] G. P. van der Kelen, *Bull. Soc. Chim. Belges* **65,** 1343 (1956).

[212] R. A. Kovar, H. Derr, D. Brandau, and J. O. Callaway, *Inorg. Chem.* **14,** 2809 (1975).

[213] G. Renwanz, *Ber.* **65,** 1308 (1932).

[214] T. Wartik and H. I. Schlesinger, *J. Am. Chem. Soc.* **75,** 835 (1953).

[215] J. J. Eisch, *J. Am. Chem. Soc.* **84,** 3605 (1962).

Lithium alkyls in excess can lead to coordination complexes of the type $LiGaR_4$, while zinc alkyls themselves are not economically feasible reagents. However, the commercial availability of a variety of inexpensive aluminum alkyls makes their use as organometallic alkylating agents most advantageous. As to a choice between a gallium metal–mercury alkyl exchange and an aluminum alkyl–gallium halide exchange, the former method requires careful and prolonged fractional distillation to separate the product from remaining mercury alkyl. On the other hand, the aluminum–gallium exchange proceeds very rapidly to completion, and selective inorganic salts can be used to complex any residual traces of aluminum compounds upon redistillation of the product.

Procedure:

$$3[(CH_3CH_2)_3Al]_2 + Ga_2Cl_6 \longrightarrow 2(CH_3CH_2)_3Ga + 3[(CH_3CH_2)_2AlCl]_2 \quad (94)$$
$$3[(CH_3CH_2)_2AlCl]_2 + 6KCl \longrightarrow 6K[(CH_3CH_2)_2AlCl_2]$$

The preparation of anhydrous gallium(III) chloride from gallium metal directly in the flask to be used for the alkylation often proves to be most convenient. Thus a 500-ml three-necked, round-bottomed flask equipped with stoppers and a three-way stopcock is charged with 13.9 g (0.20 g-atom) of gallium metal and then flushed thoroughly with dry nitrogen. The flask is evacuated under high vacuum, and then a stream of highly pure, dry chlorine gas is admitted slowly through the stopcock. The introduction of chlorine is so regulated that the exothermic reaction barely keeps the gallium chloride in the liquid state. After all traces of gallium metal have disappeared (supplemental heating may be necessary), the excess chlorine is removed by repeated evacuation and refilling with dry nitrogen.

The foregoing gallium chloride is dissolved partially in 200 ml of dry pentane, and the flask provided with a reflux condenser (Fig. 25 or copper coils), a sealed mechanical stirrer, and a pressure-equalized addition funnel (Fig. 19). Over a period of 1 hr, 82.0 ml (68.4 g, 0.60 mole) of triethylaluminum[216a] is added dropwise to the stirred slurry. After the exothermic reaction has subsided, the now clear solution is heated under reflux for another hour. The pentane is then drawn off under reduced pressure, and the viscous liquid residue is transferred to a Claisen distillation flask under a nitrogen atmosphere (Fig. 29). Then 50 g (0.67 mole) of finely ground, dried (400°) potassium chloride is introduced into the syrupy liquid. The resulting mixture is heated at 100–110° in an oil bath for 1 hr with thorough and frequent shaking. Two distinct liquid layers, the upper one of triethyl-

---

[216a] A good grade of commercially available triethylaluminum (95%+) can be used directly. Samples of uncertain origin should be redistilled (Fig. 29) to remove alkoxides (ineffective) and hydrides (causing a by-product of gallium metal).

gallium and the lower one of the complex, $K[(CH_3CH_2)_2AlCl_2]$, are formed. Distillation of the product at 300 mm Hg affords 27.9 g (89%) of triethylgallium boiling at 108–110°. The product may be freed of small amounts (<1%) of triethylaluminum by heating it for 1 hr at 100° with 3.0 g of finely powdered, ignited sodium fluoride and then distilling the triethylgallium. Recovered in >90% yield, the gallium alkyl was shown by emission spectrographic analysis of an ignited oxide sample to contain <0.1% by weight of aluminum oxide.

Triethylgallium has been prepared according to the foregoing procedure in yields of 83–90% for the singly distilled product from runs ranging between 0.1 and 0.4 mole in size. From the residue of $K[Al(C_2H_5)_2Cl_2]$ ~ 4% gallium metal can be recovered as a nugget.

The nmr spectra of triethylgallium and some complexes with Lewis bases have been analyzed in terms of chemical shift separations between the CH_2 and CH_3 groups.[216b]

21. Triisobutylgallium

The thermal instability of this β-branched alkyl at temperatures over 150° leads to decomposition into gallium metal, isobutylene, and hydrogen. Moreover, as the rate of decomposition seems to be autocatalyzed by the gallium metal formed, excessive or uneven heating should be avoided when conducting the aforementioned type of alkylation, and the diisobutylaluminum hydride content of the triisobutylaluminum should be kept low.[215]

Procedure:

$$6(\text{iso-}C_4H_9)_3Al + Ga_2Cl_6 \longrightarrow 2(\text{iso-}C_4H_9)_3Ga + 3[(\text{iso-}C_4H_9)_2AlCl]_2 \quad (95)$$
$$3[(\text{iso-}C_4H_9)_2AlCl]_2 + 6KCl \longrightarrow 6K[(\text{iso-}C_4H_9)_2AlCl_2]$$

In a manner and with a setup analogous to that described in Section E,20, 0.20 mole of gallium(III) chloride in 200 ml of dry pentane is treated with 149 ml (0.60 mole) of freshly distilled triisobutylaluminum (0.61% content of diisobutylaluminum hydride as determined by the isoquinoline method of analysis).[125,127] After subsequent heating and removal of the pentane the residue is treated with 50 g of finely ground, dried (400°) potassium chloride. The slurry is gently warmed to 90–100° for 1 hr and shaken frequently. Under reduced pressure of 3 mm Hg the upper layer of the resulting biphasic system is distilled, bp 67–69°, to provide a 44.1 g (92%) yield of triisobutylgallium (Ga 29.3%). Heating over 5 g of powdered, ignited sodium fluoride for 1 hr at 100° and redistillation give an analytically pure product.

[216b] S. Brownstein, B. C. Smith, G. Ehrlich, and A. W. Laubengayer, *J. Am. Chem. Soc.* **81**, 3826 (1959); A. Leib, M. T. Emerson, and J. P. Oliver, *Inorg. Chem.* **4**, 1825 (1965).

The nmr spectrum of neat triisobutylgallium consists of (δ) a doublet at 0.93 (CH_2), a doublet at 1.08 (CH_3), and a nine-peak multiplet at 2.23 (CH) ppm.[212]

22. Triphenylgallium

The preparation of gallium aryls can also result from the routes used for the alkyl derivatives. The Grignard approach consists in treating gallium(III) chloride with three or more equivalents of the arylmagnesium bromide, precipitating the magnesium halide with dioxane, isolating the crystalline triarylgallium dioxanate in 42–75% yield, and driving off the dioxane at 100° under vacuum over 18–20 hr to yield the gallium aryl.[217] The diaryl-mercury–gallium metal reaction has been carried out without solvent in a sealed tube at 130°.[218] In our laboratory the use of xylene has removed the need for and the dangers of a sealed tube.[219]

Procedure[219]:

$$3(C_6H_5)_2Hg + 2Ga \longrightarrow 2(C_6H_5)_3Ga + 3Hg \qquad (96)$$

In the C chamber of the dual chamber vessel depicted in Fig. 24 (chamber capacity of 150 ml) are placed 3.30 g (47.5 mg-atoms) of gallium metal (99.999%), 25.0 g (70.4 mmoles) of diphenylmercury (*caution: extremely toxic*), and 40 ml of dry xylene (distilled from sodium under nitrogen). Magnetic stirring is provided, and the system is alternately evacuated and flushed with nitrogen. The apparatus is covered with aluminum foil and tilted so that the C chamber is in an oil bath and the A chamber is above it and acts as a reflux air condenser. The reaction mixture is stirred at reflux for 7 days, and then the colorless, hot solution decanted from the mercury over into chamber A. After crystals are deposited, chamber A is cooled and the mother liquor poured into chamber C. Chamber C is replaced by another chamber (C′), containing fresh toluene. By pouring toluene onto the crystals in A and heating, the product in A is recrystallized from toluene. Chamber C or C′ then receives the mother liquor, the joint of A is stoppered, and

---

[217] I. M. Viktorova, N. I. Sherverdina, E. D. Delinskaya, and K. A. Kocheshkov, *Dokl. Acad. Nauk SSSR* **152,** 609 (1963).
[218] H. Gilman and R. G. Jones, *J. Am. Chem. Soc.* **62,** 980 (1940).
[219] J. L. Considine, B. S. Carroll, and J. J. Eisch, unpublished studies (1970).

chamber A is evacuated to remove traces of solvent. The yield of colorless crystals is ~10 g (70%), mp 166–167° (sealed tube).[220]

23. Tri-*n*-decylgallium

Although the method used for the ethyl- and isobutylgallium derivatives could be extended to longer *n*-alkyl chains by using the appropriate tri-*n*-alkylaluminums, these aluminum alkylating agents are not readily and directly accessible. Furthermore, the thermal lability of the resulting gallium alkyls puts a limit on the facile volatilization of the product from the $K[R_2AlCl_2]$ complex. Therefore a feasible alternative preparation is the olefin displacement between a 1-alkene and triisobutylgallium. Since the isobutylene and excess of alkene are drawn off, leaving a residue of the desired undistillable product, high purity of reagents, complete displacement, and adequate vacuum evaporation are essential to product purity.[215]

Procedure:

$$(\text{iso-}C_4H_9)_3Ga + 3CH_2{=}CH{-}n\text{-}C_8H_{17} \longrightarrow (n\text{-}C_8H_{17}CH_2CH_2)_3Ga + 3(CH_3)_2C{=}CH_2\uparrow \quad (97)$$

A homogeneous mixture of 11.4 g (0.047 mole) of triisobutylgallium and 26.0 g (0.185 mole) of dry 1-decene (freshly distilled over sodium) is heated at 155 ± 2° in a 100-ml two-necked flask, one neck of which is connected to a mercury-filled gas buret (Fig. 10). After 26 hr >90% of the expected amount of gas has been evolved. The colorless liquid residue is then subjected to a reduced-pressure evaporation at a bath temperature of 100° and a mercury diffusion pump pressure of 10^{-4} mm Hg. From the tared trap cooled in liquid nitrogen 6.2 g of excess decene is recovered after 90 min (calcd. excess, 6.0 g). The colorless distillation residue comprises almost a quantitative yield of analytically pure, viscous tri-*n*-decylgallium. The low residue of isobutyl groups is shown by only very minute bubble formation upon treatment with sulfuric acid.

---

[220] See comment on the purity of diphenylmercury in ref. 218. The presence of phenylmercuric halide is detrimental in these displacements.

Triphenylgallium in toluene solution undergoes photolysis with 254-nm irradiation to yield gallium metal, biphenyl, bibenzyl, and other products (J. J. Eisch and J. L. Considine, unpublished studies). Thus, to avoid the formation of finely divided metal in this preparation, light should be excluded. If metal is suspended in the final reaction solution, it can be effectively removed by inserting, between chambers C and A, a sintered-glass frit of coarse porosity upon which a layer of Celite (a type of diatomaceous earth available from Johns-Manville, Inc.) has been deposited. The hot xylene solution is then filtered through this frit into chamber A. A fresh chamber (C′) then is substituted for C.

24. Diethylindium Chloride

Organoindium halides can be prepared by (*a*) the redistribution reaction between R_3In and $InCl_3$[221]; (*b*) the partial alkylation of indium(III) halides by Grignard or organolithium reagents[222,223]; (*c*) the partial cleavage of indium alkyls by halogens[224]; and (*d*) an interesting halogen–indium exchange with haloforms.[225] Examples of the first and fourth methods are instructive.

Procedure[221]:

$$2(CH_3CH_2)_3In + InCl_3 \xrightarrow{\Delta} 3(CH_3CH_2)_2InCl \quad (98)$$

In a 100-ml Schlenk flask (Fig. 13c) a 2.87-g (25-mg-atom) sample of indium metal is chlorinated in the manner described in Section E,25. Then the flask is provided with a nitrogen inlet and a rubber septum. Over a few minutes 10.1 g (50 mmoles) of triethylindium is introduced, and the solid mass becomes quite warm. The septum is replaced by a reflux condenser, 30 ml of dry, deoxygenated pentane is added, and the mixture is heated at reflux for 30 min. Under nitrogen the gray suspension is filtered and the residue dissolved in 150 ml of hot, dry benzene. Cooling deposits 12.8 g (82%) of colorless, fine needles of diethylindium chloride, which after vacuum-drying have a melting point of 202–204° (under nitrogen).

Procedure[225]:

$$(CH_3CH_2)_3In + HCCl_3 \longrightarrow (CH_3CH_2)_2InCl + CH_3CH_2CHCl_2 \quad (99)$$

In a 50-ml Schlenk flask (Fig. 13c) are placed 2.18 g (7.9 mmoles) of triethylindium, 7.5 g (63 mmoles) of chloroform, and 10 ml of dry hexane. After a 10-min reflux period, the mixture is cooled to deposit 1.5 g (91%) of fine needles of diethylindium chloride, mp 202–205°.

25. Triethylindium

Triorganoindium compounds, in their modes of preparation and reaction, resemble their gallium counterparts rather closely. Accordingly, they can be prepared by (*a*) treating indium(III) halides with organolithium,[222] -magnesium,[223] or -aluminum[215] reagents; and (*b*) heating diorganomer-

---

[221] J. J. Eisch, unpublished studies (1956).

[222] H. C. Clark and A. L. Pickard, *J. Organometal. Chem.* **8,** 427 (1967).

[223] F. Runge, W. Zimmermann, H. Pfeifer, and D. Pfeifer, *Z. Anorg. Allgem. Chem.* **267,** 39 (1951).

[224] W. C. Schumb and H. J. Crane, *J. Am. Chem. Soc.* **60,** 306 (1938).

[225] K. Yasuda and R. Okawara, *Inorg. Nucl. Chem. Lett.* **3,** 135 (1967).

cury compounds with indium metal.[226] As a variant of the former method, the reaction of an alkyl bromide with a magnesium–indium alloy has also been successful.[227] Alkylating agents involving ethyl ether as a solvent produce triorganoindium etherates. The etherate must then be decomposed by heating to yield the unsolvated R_3In. In fact, the isolation of indium alkyls and aryls from Grignard reactions has been improved by precipitating the compound as a 1,4-dioxane complex. Heating under reduced pressure then yields the free R_3In.[223] Obviously, this method is unsuitable for thermally labile compounds, such as triisobutylindium.

The method given here yields an ether-free product. However, the addition of potassium chloride is essential for the success of the preparation. As in the preparation of gallium alkyls, this salt complexes with the diethylaluminum chloride by-product, which otherwise would complex with indium chloride and hinder the alkylation process.

Procedure[215]:

$$3[(CH_3CH_2)_3Al]_2 + In_2Cl_6 \longrightarrow 2(CH_3CH_2)_3In + 3[(CH_3CH_2)_2AlCl]_2 \quad (100)$$
$$3[(CH_3CH_2)_2AlCl]_2 + 6KCl \longrightarrow 6K[(CH_3CH_2)_2AlCl_2]$$

In a manner analogous to that described in Section E,20, 43.6 g (0.20 mole) of indium(III) chloride is prepared from indium metal and chlorine gas in a 500-ml three-necked flask. The flask is equipped as shown in Fig. 19 (without drain), and 200 ml of dry pentane is added. Then over a 45-min period is introduced 82 ml (0.60 mole) of freshly distilled triethylaluminum (low hydride content). After the vigorous exotherm has subsided (but before a white solid forms), 70 g (0.94 mole) of finely ground, ignited potassium chloride is added to the solution. For this addition, the salt is placed in a nitrogen-flushed Erlenmeyer flask and connected, via a wide rubber tube, to the neck from which the addition funnel is removed. The reaction mixture is stirred vigorously, and the pentane removed *in vacuo*. The residue is transferred to the apparatus shown in Fig. 29. The triethylindium that distills at 95–105° at 17 mm Hg amounts to 36 g (89%) but still contains 1.5% chlorine. Fractional redistillation gives a chlorine-free main fraction of 67%, bp 51–54° at 3 mm Hg.

The nmr spectrum of triethylindium shows absorption centered at (δ, CH_2Cl_2) 0.68 (q, 2H) and 1.32 (t, 3H) ppm.

26. Triisobutylindium

In applying the foregoing method[215] to the preparation of this compound, special problems are encountered. Despite the use of potassium

[226] E. Wiberg, T. Johannsen, and O. Stecher, *Z. Anorg. Allgem. Chem.* **251**, 114 (1943).
[227] E. Todt and R. Dötzer, *Z. Anorg. Allgem. Chem.* **321**, 120 (1963).

chloride, complete alkylation of the indium chloride occurs only to an extent of 50%. This observation and the observed recovery of ~16% of the starting triisobutylaluminum indicate that ~50% of the diisobutylindium chloride persists under the reaction conditions:

$$6(\text{iso-}C_4H_9)_3Al + In_2Cl_6 \longrightarrow (\text{iso-}C_4H_9)_3Al + (\text{iso-}C_4H_9)_3In + (\text{iso-}C_4H_9)_2InCl + 5(\text{iso-}C_4H_9)_2AlCl \quad (101)$$

The separation of the desired product from the volatile aluminum compounds illustrates the utility of preferential complexation. First, complexing potassium chloride with the dialkylaluminum chloride permits distillation of the aluminum and indium alkyls. Then, complexing potassium fluoride firmly with the aluminum alkyl allows isolation of the pure indium alkyl.

Procedure[215]:

$$3[(\text{iso-}C_4H_9)_3Al]_2 + In_2Cl_6 + 6KCl \longrightarrow 2(\text{iso-}C_4H_9)_3In + 6K[(\text{iso-}C_4H_9)_2AlCl_2] \quad (102a)$$

$$(\text{iso-}C_4H_9)_3Al + KF \longrightarrow K[(\text{iso-}C_4H_9)_3AlF]$$

In the same apparatus and manner as in the previous procedure, 43.6 g (0.20 mole) of indium(III) chloride in 200 ml of dry pentane is treated with 149 ml (0.60 mole) of freshly distilled (low hydride content) triisobutylaluminum over a 40-min period. After a 1-hr heating at reflux the turbid solution is freed of solvent *in vacuo* and the residue then treated with 50 g of powdered, dry potassium chloride. The suspension is shaken while being held at 110° for 30 min. Distillation gives 50 ml of an aluminum and indium alkyl mixture, bp 78–83° at 2 mm Hg. This fraction is heated with 15 g of powdered and ignited potassium fluoride for 30 min at 110°, whereupon the fluoride dissolves and two liquid layers are formed. Distillation yields 24 g of pale-yellow triisobutylindium, bp 65–66° at 2 mm Hg. The compound is light-sensitive and deposits indium metal after a time.

27. Triphenylindium

This unsolvated indium aryl can be prepared, as can its aluminum and gallium analog, from the metal and diphenylmercury.[228] (Other synthetic methods are discussed in Section E,25). To avoid the need for sealed reaction vessels, refluxing xylene can be used to provide the requisite temperature. Although not as reactive as aluminum aryls, indium and gallium aryls should be manipulated and stored in a dry, oxygen-free, preferably dark environment. Solutions of either aryl do deposit metal upon ultraviolet irradiation.[219]

As noted in Section E,15, the metal displacement reaction with diphenylmercury varies greatly in its rate, which seems to depend upon the purity

---

[228] H. Gilman and R. G. Jones, *J. Am. Chem. Soc.* **62**, 2353 (1940).

of the reagents and the state of the metal's subdivision. Prior amalgamation of the metal, with added mercury or mercuric chloride, seems to promote the reaction. In any event, a negative test for xylene-soluble mercury should determine when the heating period has been adequate.

Procedure:

$$3(C_6H_5)_2Hg + 2In \longrightarrow 2(C_6H_5)_3In + 3Hg \qquad (102b)$$

In the C chamber of the dual-chamber apparatus shown in Fig. 24 (chamber capacity of 500 ml) are placed 25.0 g (0.22 g-atom) of indium chips, 40.0 g (0.112 mole) of diphenylmercury (*caution: extremely toxic*), and 300 ml of dry xylene. Under nitrogen the mixture is heated at reflux for 2–4 days (arrangement as for triphenylgallium). After a negative soluble mercury test, the warm contents are decanted into chamber A. When suspended gray solid cannot be avoided, then the contents of C are filtered into A through a pad of Celite on a coarse glass frit. The solution in chamber A is concentrated by distilling xylene back into C and then chilled to deposit triphenylindium. The mother liquor is poured into C, and chamber C is replaced by chamber C′ containing dry toluene or ethanol-free chloroform. Recrystallization of the product in the usual manner yields 18–21 g (65–80%) of triphenylindium as colorless needles, mp 207–208°.

28. Thallium(I) Cyclopentadienide

In contrast to the air- and moisture-sensitive triorganothallium(III) compounds, this thallium(I) compound is reminiscent of the stable organomercury compounds and especially the mercury acetylides, which also can be precipitated from aqueous solution.[229–231]

Procedure:

$$2\,C_5H_6 + Tl_2SO_4 + 2KOH \longrightarrow 2\,C_5H_5Tl + K_2SO_4 + 2H_2O \qquad (103)$$

A stirred solution of 50.5 g (0.1 mole) of thallium(I) sulfate (previously recrystallized from water) and 20 g of potassium hydroxide in 400 ml of water is treated dropwise with a solution of 10 g (0.15 mole) of freshly distilled cyclopentadiene[232] in 10 ml of methanol. After vigorous stirring for 1 hr the suspension is filtered and the solid washed with cold methanol. The

---

[229] E. O. Fischer, *Angew. Chem.* **69,** 207 (1957).

[230] F. A. Cotton and L. T. Reynolds, *J. Am. Chem. Soc.* **80,** 269 (1958).

[231] J. M. Lalancette and A. Lachance, *Can. J. Chem.* **49,** 2996 (1971).

[232] See R. B. Moffett, *in* "Organic Syntheses," Coll. Vol. IV, p. 238, Wiley, New York, 1963, for the preparation of cyclopentadiene by the thermal cracking of dicyclopentadiene.

thallium(I) cyclopentadienide is dried *in vacuo* over potassium hydroxide pellets to give 20–22 g (80%) of pale-yellow crystals. The product can be sublimed between 100 and 150°.

29. Diphenylthallium Halide

Diorganothallium halides can be prepared by the action of (*a*) organomagnesium[233–235] or -lithium[236,237] reagents on thallium(III) halides, (*b*) acids on triorganothallium compounds to cleave one carbon–thallium bond, (*c*) organomercury compounds on thallium(III) halides,[238] or (*d*) arylboronic acids on thallium(III) halides in aqueous solution.[239] For diphenylthallium halides, the first method gives low yields (~20%), but the second method produces the iodide in 79% yield. Since the thallium(III) chloride[240] and bromide[241] are hygroscopic and unstable above room temperature, their preparation and use in aqueous solution are highly advantageous. Therefore their reaction with phenylboronic acid in water is the preparative method of choice for these diphenylthallium halides.

Procedure[239]:

$$2C_6H_5B(OH)_2 + TlX_3 + 2H_2O \longrightarrow (C_6H_5)_2TlX + 2B(OH)_3 + 2HX \qquad (104)$$

Caution: Thallium(III) salts and organothallium(III) derivatives are toxic. Disposable gloves should be worn, and tidy transfers of such solids should be made.

Aqueous solutions of thallium(III) bromide[241] or chloride[240] can be made by treating an aqueous suspension of thallium(I) halide with the halogen (freshly prepared bromine water or chlorine gas) and then sweeping the excess halogen out of solution with nitrogen. Alternatively, a solution of thallium(I) halide in acetonitrile (e.g., 30 ml of acetonitrile per 1.5 g of Tl) is treated with a slight excess of halogen and the cooled solution is evaporated *in vacuo*. The residue is dissolved in water.

A solution of 9.5 g (22 mmoles) of thallium(III) bromide in 50 ml of

---

[233] R. J. Meyer and A. Bertheim, *Ber.* **37,** 2051 (1904).
[234] E. Krause and A. von Grosse, *Ber.* **58,** 1933 (1925).
[235] H. Hecht, *Z. Anorg. Allgem. Chem.* **254,** 27 (1947).
[236] N. N. Melnikov and G. P. Gratscheva, *Zh. Obshch. Khim.* **6,** 634 (1936) [*Chem. Abstr.* **30,** 5557 (1936)].
[237] A. N. Nesmeyanov, A. E. Borisov, and N. V. Novikova, *Dokl. Akad. SSSR* **96,** 289 (1954) [*Chem. Abstr.* **49,** 5276 (1955)].
[238] A. E. Goddard and D. Goddard, *J. Chem. Soc.* **36,** 256, 482 (1922).
[239] F. Challenger and B. Parker, *J. Chem. Soc.* 1462 (1931).
[240] R. J. Meyer and A. Bertheim, *Ber.* **37,** 2051 (1904).
[241] (a) R. J. Meyer, *Z. Anorg. Allgem. Chem.* **24,** 321 (1900); (b) B. F. G. Johnson and R. A. Walton, *Inorg. Chem.* **5,** 49 (1966).

water is admixed with 6.0 g (50 mmoles) of phenylboronic acid, and the mixture is heated on a water bath for 5 hr. The cooled, concentrated solution deposits 7.1 g (75%). The dried diphenylthallium bromide can be recrystallized from pyridine and does not melt under 290°.

The diphenylthallium chloride can be prepared in an analogous fashion.

30. Triphenylthallium

Triorganothallium compounds can be prepared by the action of organolithium[242,243] or organomagnesium reagents on thallium(III) chloride or diorganothallium halides. For the Grignard reaction, elevated temperatures[244] or use of a THF[245] medium is necessary for complete substitution of halogen. Trialkylthallium derivatives are also accessible by the concerted action of an alkyllithium and its organic iodide on thallium(I) iodide.[246]

The disadvantage of procedures using thallium(III) chloride is the necessity for dehydrating the hygroscopic tetrahydrate. Such dehydration can be achieved in several more-or-less elaborate ways. One of the oldest ways is to dissolve the hydrated thallium chloride in absolute ether and to stir the solution with anhydrous copper(II) sulfate.[240] But the ready preparation of anhydrous diphenylthallium halides (Section E,29) makes them more accessible starting materials for the preparation of triphenylthallium.

Procedure[243]:

$$(C_6H_5)_2TlBr + C_6H_5Li \xrightarrow{Et_2O} (C_6H_5)_3Tl + LiBr \tag{105}$$

In a 500-ml Schlenk flask (Fig. 13c) equipped with a pressure-equalized funnel are placed 22.3 g (51 mmoles) of dried diphenylthallium bromide and 125 ml of anhydrous ethyl ether. The stirred suspension is cooled in an ice bath, and then 42 ml of an ethereal 1.2 *M* phenyllithium (50 mmoles) solution is added dropwise. After 30 min at 0° the ether is removed *in vacuo*, and the addition funnel replaced by a reflux condenser. A 100-ml portion of dry, deoxygenated benzene is added, and the mixture heated at reflux to dissolve the triphenylthallium. The hot suspension is poured through a coarse filter into a second Schlenk flask. Petroleum ether (bp 65–70°) is added to the filtrate to deposit fine needles of triphenylthallium, which are collected and washed with petroleum ether. Drying *in vacuo* gives 15 g (65–70%) of product, mp 169–171°.

---

[242] S. F. Birch, *J. Chem. Soc.* 1132 (1934).

[243] H. Gilman and R. G. Jones, *J. Am. Chem. Soc.* **61,** 1513 (1939); **62,** 2357 (1940).

[244] J. L. W. Pohlmann and F. E. Brinkman, *Z. Naturforsch.* **20b,** 5 (1965).

[245] O. Y. Okhlobystin, K. A. Bilevitch, and L. I. Zakharkin, *J. Organometal. Chem.* **2,** 281 (1964).

[246] H. Gilman and R. G. Jones, *J. Am. Chem. Soc.* **68,** 517 (1946); **72,** 1760 (1950).

F. Compounds of Group IVA

1. General Preparative and Analytical Methods

Unlike the organometallics of Groups I, II, and III, the tetraorgano derivatives of silicon, germanium, tin, and lead have little tendency to oxidize or hydrolyze in moist air or to form complexes with Lewis bases. These so-called organometalloids[3] require strong oxidants or highly acidic protolyzing agents to cleave the carbon–metal bond, and only organo derivatives bearing several electron-withdrawing ligands, such as $RSiF_3$, display Lewis acidic character.[247,248] They are generally unassociated, covalent compounds whose thermal stability is especially high for the silicon and germanium members and quite low for those of tin and lead. Carbon's remarkable tendency to form compounds containing chains of carbon atoms (homoconcatenation) is mirrored in the congeners of Group IVA as well. In addition to the dimetallic derivatives, R_3MMR_3, known for all these metals, a growing variety of linear and cyclic polysilanes has been synthesized and characterized over the past 25 years.[249] Moreover, Group IVA metals can also form bonds with other metals, within or outside Group IV (heteroconcatenation), as exemplified by $(C_6H_5)_3SiK$ and $[(CH_3)_3Si]_2Hg$.

These organometalloids can be prepared by an unusually broad array of methods, involving either formation of the carbon–metal bond[250] or chemical modification of one of the organic groups in R_4M. Organosilanes, for example, can be subjected to numerous reactions that leave the carbon–silicon bonds intact.[251] General ways of forming bonds between carbon and Group IVA metals in the laboratory include (*a*) reaction of organometallics of Groups I–III with the metal halide [Eq. (106)]; (*b*) reaction of organic halides with Group IVA metal metallics, R_3MM' [Eq. (107)]; (*c*) addition of Group IVA hydrides to olefins or acetylenes [Eq. (108)]; (*d*) metal–metal exchange reactions, especially between compounds of tin or lead and those

---

[247] K. Tamao *et al.*, *J. Am. Chem. Soc.* **100,** 290 (1978).

[248] K. Tamao, T. Kakui, and M. Kumada, *J. Am. Chem. Soc.* **100,** 2268 (1978).

[249] H. Gilman and G. L. Schwebke, *Adv. Organometal. Chem.* **1,** Chapter 3 (1964).

[250] C. Eaborn and R. W. Bott, *in* "Organometallic Compounds of the Group IV Elements" (A. G. MacDiarmid, ed.), Vol. I, Dekker, New York, 1968.

[251] P. F. Hudrlik, *in* "New Applications of Organometallic Reagents in Organic Synthesis" (D. Seyferth, ed.), pp. 125–160. Elsevier, Amsterdam, 1976.

of Groups I–III [Eq. (109)]; (*e*) catalyzed redistribution reaction between R_4M and MX_4 for preparing organometallic halides [Eq. (110)]; (*f*) selective metal–carbon bond cleavage [Eq. (111)]; and (*g*) chemical modification of R_nMX_{4-n} types into other halides, hydrides, alkoxides, etc. [Eq. (112)]:

$$R_3MX + R'M' \longrightarrow R_3MR' + M'X, \quad M' = Li, MgX, AlR'_2 \tag{106}$$

$$R_3MM' + R'X \longrightarrow R_3MR' + M'X, \quad M' = K, Li, Na \tag{107}$$

$$R_3MH + -C \equiv C-R' \longrightarrow (R_3M)C \overset{\cdots}{=} C(H)(R') \tag{108}$$

$$(H_2C{=}CH)_4Sn + 4C_6H_5Li \longrightarrow 4H_2C{=}CHLi + (C_6H_5)_4Sn \tag{109}$$

$$3(CH_3CH_2)_4Ge + GeBr_4 \xrightarrow{AlBr_3} 4(CH_3CH_2)_3GeBr \tag{110}$$

$$(C_6H_5)_4Pb + HCl \longrightarrow (C_6H_5)_3PbCl + C_6H_6 \tag{111}$$

$$R_3MOR' \xleftarrow[-HX]{R'OH} R_3MX \xrightarrow{LiAlH} R_3MH \tag{112}$$

Because most tetraorgano compounds of Group IVA are reasonably stable to the atmosphere and to heat, they can be purified and analyzed just like any ordinary organic compound. Decomposition with hot, concentrated sulfuric acid and ignition to the oxide is a generally acceptable assay of the metal content.[252] The light and heat sensitivity of certain tin and lead derivatives and the great hydrolyzing tendency of halides, hydrides, and alkoxides should be borne in mind during such separations and analyses. Finally, the generally toxic character of lead compounds should be taken into careful account.[253]

2. 1,1-Dimethyl-2,5-dihydrobenzo[*c*]silole

Metallocyclic compounds are really only a special class of unsymmetric organometallic derivatives of the type R_2MR'. Accordingly, they can be synthesized from α,ω-dihalo or α,ω-dimetallo organic derivatives (R_2 = α,ω-disubstituted group) and either a dimetallic or dihalogen derivative of M, namely, $R'MM'_2$ or $R'MX_2$:

$$C_n\begin{matrix}X\\X\end{matrix} + \begin{matrix}M'\\M'\end{matrix}M \longrightarrow \widehat{C_n\ \ M} + 2M'X \quad (\text{or } C_nM'_2 + MX_2) \tag{113}$$

In the synthesis of many silicon heterocycles, R_2SiX_2 is simply allowed to react with α,ω-dilithio or α,ω-dimagnesio organic systems. As reaction partners in such reactions, allylic and benzylic metallic derivatives often give poor results. Attempts to generate these RM types from a metal and the halide, RX, lead preferentially to the coupling of RM and RX to yield R—R. This difficulty can be reduced by the classic technique introduced by

---

[252] H. Gilman and L. S. Miller, *J. Am. Chem. Soc.* **73,** 968 (1951).

[253] H. Shapiro and F. W. Frey, "The Organic Compounds of Lead," Wiley, New York, 1968.

Barbier[254]: generating the reactive organometallic in the presence of a reactive third component which reacts more readily with RM than does RX. The synthesis of the following silole illustrates the value of this approach, for α,α′-dibromo-*o*-xylene cannot easily be converted into its α,α′-dimetallo derivative. Treatment with an active metal produces, instead, dibenzo-[*a,e*]1,5-cyclooctadiene and its oligomers.[255]

Procedure[256]:

$$o\text{-}C_6H_4(CH_2Br)_2 \xrightarrow[-LiBr]{2Li} [o\text{-}C_6H_4(CH_2Li)(CH_2Br)] \xrightarrow[-LiCl]{(CH_3)_2SiCl_2} o\text{-}C_6H_4(CH_2Br)(CH_2Si(CH_3)_2Cl) \xrightarrow[-2LiX]{2Li} o\text{-}C_6H_4(CH_2)_2Si(CH_3)_2 \quad (114)$$

Caution: α,α′-Dibromo-o-xylene is a powerful lachrymator and skin irritant. Disposable gloves should be worn, and all transfers and manipulations should be carried out in an efficient hood. The chemist should remember not to bring the gloved hands anywhere near the face (such as unconsciously scratching or adjusting goggles). Dichloro(dimethyl)silane is volatile, easily hydrolyzed, and also *highly irritating*.

A 1-liter three-necked flask is equipped as in Fig. 19, except that a low-temperature thermometer dipping into the liquid replaces the reflux condenser. After 200 ml of anhydrous ethyl ether and 14.5 g (84 mmoles) of dichloro(dimethyl)silane are placed in the flask, the thermometer is replaced by a solids funnel and, in an emerging current of nitrogen, 4.1 g (0.6 g-atom) of lithium wire is cut with scissors into 2- to 3-mm pieces and dropped directly into the flask. The addition funnel is charged with 20.1 g (76 mmoles) of α,α′-dibromo-*o*-xylene and 200 ml of dry ether, and the contents of the flask cooled to −60 to −70°. During 75 min the dibromide solution is added dropwise with rapid stirring to the cooled solution. A sign of good initiation and continuation of reaction is a very shiny lithium surface. The addition funnel is rinsed several times with ether, and the extracts added to the flask to remove any crystals of dibromide. After ~8 hr the cooling bath has thawed

---

[254] P. Barbier, *C. R. Acad. Sci. Paris* **128,** 110 (1899).

[255] E. Müller and G. Röscheisen, *Chem. Ber.* **90,** 543 (1957).

[256] This procedure was devised by M. M. Robillard, The Catholic University of America, 1968, and checked by A. Santolo, A. Mason, and R. J. Manfre, The State University of New York at Binghamton, 1977–1978.

and room temperature is attained. The solution is drained from the white precipitate and excess lithium. The precipitate is washed with benzene, and then the combined organic extracts are evaporated to a small volume. (Refiltration from some solid may be necessary.) Distillation at reduced pressure gives ~15 ml between 35 and 50° at 1 mm Hg. Redistillation yields 5.6 g (46%), bp 42–44° at 0.1 mm Hg, n_D^{22} 1.5232.[257] The reaction gives yields of 35–50% on scales up to thrice the foregoing.

The nmr spectrum of this benzosilole shows absorptions at (δ, external TMS, neat) 0.45 (s, 6H), 2.20 (s, 4H), and 6.85 (sym, m, 4H) ppm.

3. Trimethyl(phenylethynyl)silane

Alkynylsilanes, like most unsymmetric silanes, can be prepared in high yield from the reaction of alkynylmagnesium or -lithium reagents with chlorosilanes. These necessary alkynylmetallics can be readily obtained by the metalation of terminal acetylenes by alkyl Grignard or lithium reagents. The use of methyl or ethyl metallics has the advantage of leading to a volatile gas whose evolution serves to mark the course of conversion to the 1-alkynyl salt.[185]

Procedure[185]:

$$C_6H_5{-}C{\equiv}C{-}H + CH_3CH_2MgBr \longrightarrow C_6H_5{-}C{\equiv}C{-}MgBr + CH_3CH_3\uparrow \quad (115)$$
$$C_6H_5{-}C{\equiv}C{-}MgBr + (CH_3)_3SiCl \longrightarrow C_6H_5{-}C{\equiv}C{-}Si(CH_3)_3 + MgBrCl$$

In a 1-liter three-necked flask equipped as in Fig. 19, but without the filter drain, is prepared a solution of ethylmagnesium bromide from 55 g (0.5 mole) of ethyl bromide and 12.2 g (0.5 g-atom) of magnesium turnings in 250 ml of dry ethyl ether. Then a solution of 51 g (0.5 mole) of freshly distilled phenylacetylene in 50 ml of ether is added dropwise with stirring. The solution can be stirred overnight at 25° or heated at reflux until gas evolution ceases. Finally, a solution of 55 g (0.51 mole) of chloro(trimethyl)-silane (*caution: easily hydrolyzed and extremely irritating*) in 50 ml of dry ether is added over a 4-hr period. Magnesium halide precipitates immediately, and effective blade stirring is required to manage the semisolid mass. After 24 hr at reflux, the cooled reaction mixture is cautiously treated with 100 ml of saturated, aqueous ammonium chloride solution (inert atmosphere no longer necessary). The separated organic layer is dried over anhydrous calcium sulfate, the ether evaporated, and the residue distilled to give 69.2 g (80%) of trimethyl(phenylethynyl)silane, bp 46–47°, at 0.4 mm Hg, n_D^{22} 1.5272.

The foregoing has been adapted for the preparation of trimethylsilyl

[257] N. S. Nametkin and V. M. Vdovin, *Dokl. Akad. Nauk SSSR* **154,** 383 (1964).

derivatives from less acidic and more volatile alkynes. In these cases, it is advantageous to provide the apparatus with a cold-finger condenser charged with acetone and solid carbon dioxide and to use an excess of acetylene. In this manner, propyne, *tert*-butylacetylene, and 2-methylbut-1-en-3-yne can be converted into their trimethylsilyl derivatives in ~50% yield.

4. Trimethyl(*cis*-β-styryl)silane and Trimethyl(*trans*-β-styryl)silane

Unsymmetric vinylic silanes can be prepared by (*a*) the coupling of vinylic metallic reagents with organosilyl halides, (*b*) the hydrosilylation of acetylenes,[258] (*c*) the reaction of vinylsilyl halides with organomagnesium or -lithium reagents,[259] (*d*) the coupling of organosilylmetallic agents with vinylic halides, and (*e*) the reduction of alkynylsilanes.[185] The last method is both convenient and versatile, since the reduction with R_2AlH agents can be directed to yield a preponderance (>95%) of either the cis or the trans isomer.

Procedure[185]:

$$(C_6H_5)(H)C{=}C(H)(Si(CH_3)_3) \xleftarrow[2.\ H_2O]{1.\ R_2AlH} C_6H_5C{\equiv}C{-}Si(CH_3)_3 \xrightarrow[2.\ H_2O]{1.\ R_2AlH\cdot R'_3N} (C_6H_5)(H)C{=}C(Si(CH_3)_3)(H) \quad (116)$$

The 200-ml three-necked apparatus shown in Fig. 19, but without the filter drain, is charged with 9.8 g (56 mmoles) of trimethyl(phenylethynyl)-silane and 20 ml of dry heptane. After thorough degassing and flushing with nitrogen, the addition funnel is charged with 40 ml of heptane and 8.0 g (56 mmoles) of redistilled diisobutylaluminum hydride (>96% pure). Then, with a syringe, 4.8 g (56 mmoles) of anhydrous, degassed *N*-methylpyrrolidine is slowly added to the hydride solution (slight exotherm) and the resulting solution added dropwise to the silane. The reaction mixture is heated at 55–67° for 3 hr, cooled, and cautiously treated with 3.1 ml of water (inert cover discontinued). Filtration, solvent removal from the filtrate, and distillation of the residue yield 9.4 g (96%) of trimethyl(*cis*-β-styryl)silane, bp 50–52° at 0.2 mm Hg, whose gas chromatogram shows ~4% trans isomer. Fractional distillation leads to the pure cis isomer.

The nmr spectrum of the cis isomer shows absorptions at (δ, neat) 0.03 (s, 9H), 5.83 (d, 1H, $J = 15$ Hz), 7.2 (s, sh, 5H), and 7.36 (d, 1H, $J = 15$ Hz) ppm.

The pure trans isomer can be prepared by following the procedure given

---

[258] E. Y. Lukevits and M. G. Voronkov, "Organic Insertion Reactions of Group IV Elements," Consultants Bureau, New York, 1966.

[259] D. Seyferth, *Prog. Inorg. Chem.* **3**, 129–280 (1962).

for diisobutyl[(*Z*)-1-trimethylsilyl-2-phenylethenyl]aluminum. This product is cautiously hydrolyzed with 7.6 ml of water and worked up as given for the cis isomer. Distillation yields 23.5 g (96%) of trimethyl(*trans*-β-styryl)-silane, bp 90–91° at 13 mm Hg, whose gas chromatogram shows <5% cis isomer and starting silane. The nmr spectrum of the trans isomer shows absorptions at (δ, neat) 0.14 (s, 9H), 6.33 (d, 1H, $J = 19$ Hz), 6.88 (d, 1H, $J = 19$ Hz), and 7.07–7.41 (m, 5H) ppm.

5. Triphenylsilylpotassium and Triphenylsilyllithium

These organoidal metallic compounds can be readily prepared by cleaving either disilanes or chlorosilanes with alkali metals. With chlorosilanes, the reaction proceeds via the disilane:

$$R_3SiCl \xrightarrow[-MCl]{2M} R_3SiM \xrightarrow[-MCl]{R_3SiCl} R_3Si{-}SiR_3 \xrightarrow{2M} 2R_3SiM \tag{117}$$

Cleavage of the disilane requires a donor solvent. Therefore in the absence of a donor chlorosilanes give disilanes. In the presence of ethers, chlorosilanes proceed on to give silylmetallics. The following synthesis of triphenylsilyllithium from chloro(triphenyl)silane exemplifies this direct approach. Yet the synthesis of this silylmetallic from hexaphenyldisilane has the advantages of giving no metallic salt by-product and of employing a stable, readily storable starting silane. The tendency of chlorosilanes to hydrolyze often makes stored samples unsuitable for these reactions.[260,261]

Procedure[262–264]

$$\begin{gathered}(C_6H_5)_3Si{-}Si(C_6H_5)_3 + 2K \xrightarrow{Et_2O} 2(C_6H_5)_3SiK \\ (C_6H_5)_3SiK + (CH_3)_3SiCl \longrightarrow (CH_3)_3Si{-}Si(C_6H_5)_3 + KCl\end{gathered} \tag{118}$$

Caution: Sodium–potassium alloy should be stored and transferred with the exclusion of moisture or any active-hydrogen compound. It hydrolyzes explosively and can cause wet organic solvents to ignite.

A 250-ml three-necked flask is fitted with a motor-driven stirrer whose blade fits the contour of the bottom and almost scrapes the wall of the flask. The side necks bear a nitrogen inlet and glass stopper, respectively. Then 5.2 g (10 mmoles) of pure hexaphenyldisilane (mp >360°), 1.3 ml (transfer cautiously by pipet) of sodium–potassium alloy, and 5 ml of anhydrous ether are added, and the yellowish suspension stirred thoroughly for 2 hr.

[260] D. Wittenberg and H. Gilman, *Q. Rev.* **13,** 116 (1959).

[261] H. Gilman and H. J. S. Winkler, *in* "Organometallic Chemistry" (H. Zeiss, ed.), pp. 270–345. Van Nostrand-Reinhold, Princeton, New Jersey, 1960.

[262] R. A. Benkeser and R. G. Severson, *J. Am. Chem. Soc.* **73,** 1424 (1951).

[263] H. Gilman and T. C. Wu, *J. Am. Chem. Soc.* **73,** 4031 (1951).

[264] H. Gilman and T. C. Wu, *J. Org. Chem.* **18,** 753 (1953).

The khaki-colored paste is diluted with 50 ml of dry ether and stirred for 22 hr. Almost a quantitative yield of triphenylsilylpotassium is present at this point, as derivatization shows.

Then a solution of 2.2 g (20 mmoles) of redistilled chloro(trimethyl)silane in 30 ml of dry ether is introduced, and the turbid, colorless mixture stirred for 30 min. *Caution: Before proceeding further, a check should be made to see that no significant amounts of alloy are suspended in the mixture.* Then 95% ethanol is cautiously added dropwise to destroy any alkali metal, cooling the contents and maintaining the nitrogen atmosphere. Thereafter water is slowly added. The ether layer is separated and dried over anhydrous sodium sulfate. Evaporation of the ether leaves 6.2 g (94%) of 1,1,1-trimethyl-2,2,2-triphenyldisilane, mp 101–107°. Recrystallization from 95% ethanol gives fine, colorless needles, mp 109–110°, 5.2 g (78%).

Procedure[265]*:*

$$(C_6H_5)_3SiCl + 2Li \xrightarrow{THF} (C_6H_5)_3SiLi + LiCl \quad (119)$$

In a 200-ml three-necked flask equipped as in Fig. 19 are placed 5.9 g (20 mmoles) of pure chloro(triphenyl)silane and 550 mg (80 mg-atoms) of lithium wire pieces (cut freshly into 2-mm segments which are allowed to fall directly into the nitrogen stream emerging from the flask). To the rapidly stirred mixture is slowly added 50 ml of anhydrous THF. After a 5-min induction, the reaction begins and over 4 hr at 25° the mixture turns a clear dark brown. The triphenylsilyllithium solution is filtered into a nitrogen-flushed Schlenk flask (Fig. 13c), and 2.8 g (25 mmoles) of freshly distilled chloro(trimethyl)silane is added by syringe. Workup according to the previous procedure gives 5.6 g (85%) of 1,1,1-trimethyl-2,2,2-triphenyldisilane, mp 108–110°, after recrystallization.

6. Hexaphenyldisilane

The preparation of symmetrically substituted disilanes can be most readily achieved by the dehalogenative coupling of triorganosilyl chlorides by active metals. The alternative method, namely, the coupling

$$\begin{aligned} R_3SiCl + 2M &\longrightarrow R_3Si{-}SiR_3 + 2MCl \\ R_3SiM + R'_3SiCl &\longrightarrow R_3Si{-}SiR'_3 + MCl \end{aligned} \quad (120)$$

of triorganosilylmetallics and triorganosilyl chlorides, is also applicable in the synthesis of unsymmetrically substituted disilanes. The foregoing procedure for the preparation of triphenylsilylmetallics illustrates this approach.[260,261]

[265] M. V. George, D. J. Peterson, and H. Gilman, *J. Am. Chem. Soc.* **82,** 403 (1960).

Procedure[264]*:*

In a 1-liter three-necked flask equipped as in Fig. 19 (without drain or addition funnel) are placed 400 ml of dry xylene and 15 g (0.65 g-atom) of freshly cut sodium pieces. The mixture is stirred at reflux to form a sodium dispersion, cooled to 80°, and then treated slowly with 180 g (0.61 mole) of pure chloro(triphenyl)silane (*it is important that the melting point be sharp*, mp 95–97°). The mixture is heated at reflux and stirred such that the blade keeps any crust of solid from forming on the bottom of the flask. After a 6-hr reflux period the suspension is cooled and filtered. The filter cake is then added, slowly and in small portions, to 500 ml of 95% ethanol in order to destroy the residue sodium. (*Caution: If a poor grade of chloride is used or stirring is ineffectual, much sodium may remain.*) The solid disilane is filtered off and washed thoroughly, twice with 20% aqueous ethanol and once with ether. Drying to constant weight at 110° gives 127 g (80%), mp 364–365°.

7. 1,4-Bis(triphenylsilyl)butane

The tendency of α,β-unsaturated organosilanes and -germanes to form radical anions can be utilized for forming dimeric or cyclic derivatives[266]:

$$R_2M(\text{cyclohexane ring}) \longleftarrow R_2M(CH{=}CH_2)_2 \longrightarrow R_2M(CH{=}CH_2)CH_2CH_2CH_2CH_2M(CH{=}CH_2)R_2 \qquad (121)$$

In order to suppress competing anionic polymerization of the vinyl metalloid, it is often advantageous to conduct the coupling reaction in the presence of a weak proton donor, such as *tert*-butyl alcohol. In other cases, such as the coupling of phenylsilanes to yield 4,4′-bistriorganosilyl-1,1′,4,4′-tetrahydrobiphenyls, the resulting alkali metal adduct can be generated in high yield.

Procedure[267]*:*

$$2(C_6H_5)_3Si{-}CH{=}CH_2 \xrightarrow[2.\ H_2O]{1.\ Li,\ THF} (C_6H_5)_3Si{-}CH_2CH_2CH_2CH_2{-}Si(C_6H_5)_3 \qquad (122)$$

A 1-liter three-necked flask equipped as in Fig. 19 (without drain and with a low-temperature thermometer in place of the addition funnel) is charged with 14.3 g (50 mmoles) of triphenyl(vinyl)silane and 250 ml of purified, anhydrous THF. In an emerging current of nitrogen, 120 mg of lithium wire (6% sodium content) is cut finely and the pieces allowed to drop into the flask. Upon stirring the lithium becomes coated with a yellow

[266] J. J. Eisch and G. Gupta, *J. Organometal. Chem.* **168,** 139 (1979).
[267] J. J. Eisch and R. J. Beuhler, *J. Org. Chem.* **28,** 2876 (1963).

layer. The stirred contents are then cooled to −70 to −75° in an acetone–solid carbon dioxide bath. After 1.5 hr a second 120-mg portion of lithium is introduced in the same manner and, after a further 2 hr, the last 120-mg portion of lithium [total of 360 mg (52 mg-atoms)]. After an additional 2-hr stirring the yellow reaction suspension is diluted with 150 ml of THF (deoxygenated and precooled to −75°). After 4 hr more of stirring at −75° the suspension is cautiously treated with 200 ml of 1 *N* sulfuric acid. After warming to room temperature the mixture is diluted with 200 ml of ether. The separated organic layer is washed with aqueous sodium bicarbonate and with water. After drying over anhydrous sodium sulfate the organic layer is freed of solvent to yield 13.7 g of residue. The crude product is extracted with 150 ml of hot benzene, and the filtered extract cooled to yield 10.7 g (75%) of 1,4-bis(triphenylsilyl)butane, mp 210–213°. Recrystallization from cyclohexane yields fine needles, mp 214–215°.

8. Bromo(triphenyl)germane and Chloro(triphenyl)germane

Organogermyl halides can be prepared from R_4Ge compounds by cleavage with halogens or hydrogen halides,[268,269] or by an aluminum halide-catalyzed redistribution reaction with GeX_4.[270] In cleavage reactions, bromine is the most convenient halogen group. The bromide can be hydrolyzed to the digermyl oxide, and the latter converted by other hydrohalic acids into related halides:

$$R_3GeBr \xrightarrow{NaOH} R_3Ge{-}O{-}GeR_3 \xrightarrow{HCl} R_3GeCl \qquad (123)$$

Formation of the digermyl oxide can also be used as an aid in isolating germyl halides from the incomplete alkylation of $GeCl_4$ by Grignard reagents.[271,272] This approach is illustrated in the preparation of chloro(triphenyl)germane.[273]

Procedure[268]:

$$(C_6H_5)_4Ge + Br_2 \xrightarrow[\Delta]{BrCH_2CH_2Br} (C_6H_5)_3GeBr + C_6H_5Br \qquad (124)$$

Caution: Although not sensitive to oxidation, germyl halides are hydrolyzed in moist air, hence are skin irritants. A dry medium should be maintained for their reaction and purification.

The gas inlet of a 1-liter three-necked flask equipped as in Fig. 19 (without

---

[268] O. H. Johnson and D. M. Harris, *J. Am. Chem. Soc.* **72,** 5566 (1950).
[269] C. A. Kraus and E. A. Flood, *J. Am. Chem. Soc.* **54,** 1635 (1932).
[270] P. Mazerolles, *C. R. Acad. Sci. Paris* **251,** 2041 (1960).
[271] C. A. Kraus and L. S. Foster, *J. Am. Chem. Soc.* **49,** 457 (1927).
[272] C. A. Kraus and C. R. Wooster, *J. Am. Chem. Soc.* **52,** 372 (1930).
[273] J. J. Eisch, unpublished studies.

drain) is connected to a calcium chloride drying tube. The flask is charged with 76 g (0.2 mole) of tetraphenylgermane and 500 ml of dry, distilled 1,2-dibromoethane. To the stirred, gently boiling solution is added 35 g (0.22 mole) of bromine in a dropwise manner. Reflux is continued until the bromine vapor has almost disappeared (90 min). Then the volatiles are removed under reduced pressure. The residual solid is recrystallized once from petroleum ether, bp 40–60°, and twice from glacial acetic acid to yield 54 g (70%) of bromo(triphenyl)germane as thick needles, mp 138–139°.

Procedure[273]:

$$3C_6H_5MgBr + GeCl_4 \xrightarrow{Et_2O} (C_6H_5)_3GeX + 3MgX_2 \quad (125)$$
$$2(C_6H_5)_3GeX + 2KOH \xrightarrow{\Delta} (C_6H_5)_3Ge{-}O{-}Ge(C_6H_5)_3 + 2KX$$
$$(C_6H_5)_3Ge{-}O{-}Ge(C_6H_5)_3 + 2HCl \longrightarrow 2(C_6H_5)_3GeCl + H_2O$$

As in the preparation of *phenylmercuric bromide* (Section D,6), phenylmagnesium bromide is prepared from 157 g (1.0 mole) of bromobenzene and 24.3 g (1.0 g-atom) of magnesium turnings in 560 ml of ethereal solution (0.97 mole by titration). Then a 2-liter three-necked flask equipped as in Fig. 19 (without drain and with 1000-ml addition funnel) is charged with 800 ml of anhydrous ethyl ether and 63 g (0.294 mole) of germanium(IV) chloride. The contents are stirred in an ice bath while the Grignard solution is added over a 2-hr period. The thick white slurry is stirred for 15 hr, whereupon the Gilman color test I is negative. The mixture is then slowly poured into a large beaker containing 250 ml of concentrated hydrochloric acid and 300 g of ice. After thawing the yellow ether layer is separated and dried over anhydrous sodium sulfate. The solvent is removed, and the residue is heated at reflux for 4 hr with a solution of 25 g of potassium hydroxide in 500 ml of absolute ethanol. Dilution of the ethanol with six times its volume of water precipitates a gelatinous solid. This solid is collected and heated at reflux for 1 hr with a mixture of 450 ml of absolute ethanol, animal charcoal, and 75 ml of concentrated hydrochloric acid. Cooling deposits 41 g (41%) of chloro(triphenyl)germane as colorless needles, mp 110–115°. This product is sufficiently pure for most purposes, as the preparation of *p*-bromophenyl(triphenyl)germane (Section B,3) shows. Recrystallization from petroleum ether (bp 40–60°) produces a mp of 114–115°.

9. Triphenyl(vinyl)germane

The feasible preparation of vinyl Grignard reagents from vinyl halides in THF has made possible the synthesis of many vinyl derivatives of Group II–V metals or metalloids. The vinyl group is a most useful ligand for exploring the physical and chemical consequences of p_π-p_π or d_π-p_π bonding between the metal and unsaturated carbon. Usually the metal halide or alkoxide is treated with the appropriate ratio of the vinyl Grignard reagent[274]:

$$(CH_3)_2Si(CH{=}CH_2)_2 \xleftarrow{\frac{1}{2}(CH_3)_2SiCl_2} CH_2{=}CHMgX \xrightarrow[2.\ H_2O]{1.\ B(OCH_3)_3} CH_2{=}CHB(OH)_2 \quad (126a)$$

In the present example, the unsymmetric tetraorganogermane, triphenyl(vinyl)germane, can be readily prepared and isolated by the Grignard method. Such derivatives show no tendency to disproportionate at temperatures under 150°.[275] For other vinyl derivatives, however, such rearrangement makes isolation difficult. Dimethyl(vinyl)borane, for example, tends to rearrange to trivinyl- and trimethylboranes when stored at room temperature.[276]

Procedure[275]:

$$(C_6H_5)_3GeBr + CH_2{=}CHMgBr \longrightarrow (C_6H_5)_3GeCH{=}CH_2 + MgBr_2 \quad (126b)$$

In a 250-ml three-necked flask (Fig. 19) are placed 30 ml of anhydrous THF and 2.8 g (0.115 g-atom) of magnesium turnings. After initiation of the reaction, a solution of 8.3 ml (0.12 mole) of vinyl bromide in 70 ml of THF is slowly introduced. The resulting solution is then treated dropwise with a solution of 19.2 g (0.05 mole) of bromo(triphenyl)germane in 50 ml of dry benzene. The reaction mixture is heated at reflux for 2 hr and then hydrolyzed with aqueous ammonium chloride. The separated organic layer is dried over anhydrous sodium sulfate, the solvent evaporated, and the residue recrystallized from 95% ethanol. The triphenyl(vinyl)germane melts at 63–64° (13.5 g, 82%).

10. Tetraphenylgermane

Symmetric tetraorganogermanes can be prepared by the reaction of germanium(IV) chloride with organozinc,[277] -aluminum,[278] -magnesium,[279] and -lithium[280] reagents. The latter two reagents are the more generally suitable, and the last reagent gives the highest yields. These organogermanium derivatives are the most versatile starting materials for other germanes, such as the halides (halogenative cleavage), the digermanes

---

[274] H. D. Kaesz and F. G. A. Stone, *in* "Organometallic Chemistry" (H. Zeiss, ed.), pp. 88–149. Van Nostrand-Reinhold, Princeton, New Jersey, 1960.

[275] M. C. Henry and J. G. Noltes, *J. Am. Chem. Soc.* **82,** 555 (1960).

[276] T. D. Parsons and D. M. Ritter, *J. Am. Chem. Soc.* **76,** 1710 (1954).

[277] C. Winkler, *J. Prakt. Chem.* [2] **36,** 177 (1887).

[278] F. Glockling and J. R. C. Light, *J. Chem. Soc. A* 623 (1967).

[279] O. H. Johnson, *Inorg. Syn.* **5,** 64 (1957).

[280] O. H. Johnson and W. H. Nebergall, *J. Am. Chem. Soc.* **71,** 1720 (1949).

(Wurtz coupling of halogermanes), and unsymmetric tetraorganogermanes (treatment of R_nGeX_{4-n} with R′MgX or R′Li).

$$R_4Ge + X_2 \text{ (or HX)} \longrightarrow R_3GeX \longrightarrow R_2GeX_2 \quad (127)$$

$$R_3GeR' \xleftarrow{R'M} R_3GeX \xrightarrow{M} R_3Ge—GeR_3 \quad (128)$$

Procedure[280]*:*

$$4C_6H_5Li + GeCl_4 \xrightarrow[\Delta]{C_6H_5CH_3} (C_6H_5)_4Ge + 4LiCl \quad (129)$$

As in the preparation of tetraphenyllead, a solution of 0.75–0.77 mole of phenyllithium is prepared and placed in an addition funnel. A 2-liter three-necked flask equipped as in Fig. 19 (without drain) is charged with 10.5 g (0.05 mole) of germanium(IV) chloride and 200 ml of anhydrous ethyl ether. The stirred solution is cooled with an ice bath and treated with the phenyllithium solution over a 75-min period. Then, under a nitrogen atmosphere, the reflux condenser is replaced by an adapter connected to a downward condenser and, while toluene is slowly introduced, the ether is distilled off. When the internal temperature of the mixture exceeds 100°, the condenser is rearranged for reflux and the mixture heated at reflux for 5 hr. The cooled mixture is slowly poured into cold, aqueous 1 *N* hydrochloric acid, the organic layer separated (any solid product is filtered off) and dried over anhydrous sodium sulfate, and the solvent removed *in vacuo* to leave 9.5 g (88%) of tetraphenylgermane, mp 228–230°. From ligroin, colorless, stout needles are formed, mp 230–231°.

From the interaction of 0.1 mole of germanium(IV) chloride with 0.9 mole of phenylmagnesium bromide, also in refluxing toluene, a 73% yield of product has been reported.[279]

11. Hexamethylditin

The synthesis of unsymmetrically substituted dimetallic derivatives, $R_mM—MR'_m$, must be done by controlled coupling of the dissimilar R_mM groups. This is usually achieved by the elimination of A and B from R_mMA and R'_mMB, where A = alkali metal or hydrogen and B = halogen, alkoxyl, or amino groups. The following syntheses are illustrative[281]:

$$(C_6H_5)_3SiLi + (CH_3)_3SiCl \longrightarrow (C_6H_5)_3Si—Si(CH_3)_3 + LiCl \quad (130)$$

$$(CH_3CH_2)_3SnH + (CH_3)_3SnN(CH_2CH_3)_2 \longrightarrow (CH_3CH_2)_3Sn—Sn(CH_3)_3 + (CH_3CH_2)_2NH \quad (131)$$

Symmetrically substituted dimetallic derivatives, $R_mM—MR_m$, on the other hand, can be readily prepared in a Wurtz fashion,[282] i.e., by the

[281] W. P. Neumann, B. Schneider, and R. Sommer, *Justus Liebigs Ann. Chem.* **692,** 1 (1969).
[282] C. A. Kraus and W. V. Sessions, *J. Am. Chem. Soc.* **48,** 2361 (1926).

dehalogenative coupling of R_mMX, in either liquid ammonia or organic donor solvents.

Procedure:

$$2(CH_3)_3SnCl + 2Na \xrightarrow{\text{liq. } NH_3} (CH_3)_3Sn\text{—}Sn(CH_3)_3 + 2NaCl \quad (132)$$

A 1-liter three-necked flask equipped with a stirrer, a low-temperature thermometer, and a Dewar condenser packed with an acetone–solid carbon dioxide slurry is cooled in a crock of mineral oil to which pieces of solid carbon dioxide are added periodically. Then 300 ml of liquid ammonia is condensed into the flask, and 19.9 g (0.1 mole) of chloro(trimethyl)tin added. With stirring at -35 to $-40°$, small pieces of 2.3 g (0.1 g-atom) of freshly sliced sodium are added periodically. The color of dissolved sodium disappears, and the ditin forms a colorless precipitate. After all the sodium has reacted, the suspension is allowed to settle. Then, by siphoning or pipetting, as much of the supernatant ammonia as possible is removed and allowed to evaporate separately. The residual solid is allowed to attain room temperature, and all subsequent operations are performed under a nitrogen atmosphere. (Evaporation of the original ammonia suspension would entail some loss of the volatile ditin.) After cautious addition of 100 ml of water to the solid residue, the product is extracted into ether, the ether extracts dried over anhydrous sodium sulfate, and the solvent removed. Distillation of the product under a nitrogen atmosphere gives 13 g (80%) of colorless hexamethylditin, bp 69–70° at 20 mm Hg, mp 22–23°, d^{25} 1.570. The nmr spectrum in benzene solution shows a singlet at (δ) 0.23 ppm.

12. Di-*n*-butylin Dihydride

Tin hydrides can generally be prepared by (*a*) reduction of the corresponding halides or alkoxides by metal hydrides, and (*b*) protolysis of the organoidal metallic compound:

$$R_nSnE_{4-n} + MH_{4-n} \longrightarrow R_nSnH_{4-n} + ME_{4-n} \qquad E = X, OR \quad (133)$$

$$R_nSnM_{4-n} + (4-n)H\text{—}Z \longrightarrow R_nMH_{4-n} + (4-n)M'Z \quad (134)$$

For preparation of stannanes or tin hydrides by the first method, R_2AlH,[283] B_2H_6,[284] $LiAlH_4$,[285] and $NaBH_4$[286] have been successful. The second method has been applied to tin–sodium compounds generated in liquid

[283] W. P. Neumann, *Angew. Chem.* **75,** 225 (1963).

[284] E. Amberger and M. R. Kula, *Chem. Ber.* **96,** 2560 (1963).

[285] G. J. M. van der Kerk, J. G. Noltes, and J. G. A. Luijten, *J. Appl. Chem.* **7,** 366 (1957).

[286] M. R. Kula, E. Amberger, and H. Rupprecht, *Chem. Ber.* **98,** 629 (1965).

ammonia; in this situation, ammonium bromide proves to be a suitable proton source.[287]

These tin hydrides can be stored in the dark under nitrogen. Although not extremely air-sensitive, they do oxidize and decompose into R_4Sn, tin, and hydrogen.

Procedure[285]:

$$2(n\text{-}C_4H_9)_2SnCl_2 + LiAlH_4 \longrightarrow 2(n\text{-}C_4H_9)_2SnH_2 + LiCl + AlCl_3 \quad (135)$$

In a 1.5-liter, three-necked flask equipped as shown in Fig. 19 are placed 500 ml of anhydrous ethyl ether and 9.1 g (0.24 mole) of powdered lithium aluminum hydride. The stirred suspension is treated dropwise with a solution of 125 g (0.412 mole) of di-*n*-butyl(dichloro)tin in 250 ml of ether. The mixture is heated at reflux for 4 hr and then stirred at room temperature for 4 hr more. Then, very slowly and with cooling, the mixture is treated with 100 ml of saturated sodium sulfate solution. (*Caution: Heat and hydrogen are evolved.*) The ether layer is dried over anhydrous sodium sulfate and evaporated. The residue is distilled under a nitrogen atmosphere to give 84 g (87%) of di-*n*-butylin dihydride, bp 50–54° at 1 mm Hg.

The nmr spectrum of di-*n*-butyltin dihydride shows absorptions at (δ, neat) 0.6–1.7 (m, 18H) and 4.48 (q. 2H, $J = 2$ Hz) ppm.

13. Trimethyl(phenylethynyl)tin

Although the phenylethynyl derivatives of silanes, germanes, and stannanes can, in principle and in practice, be obtained by treating the magnesium or lithium salt of phenylacetylene with the appropriate metalloidal halide,[185] the tin analog require special consideration. While the silicon and germanium acetylides are rather stable to hydrolysis, the tin acetylides are sensitive to basic, and especially acidic, cleavage. Hence a nonhydrolytic isolation is preferable.[288]

Procedure[185,288]:

$$C_6H_5{-}C{\equiv}C{-}MgBr + (CH_3)_3SnCl \longrightarrow (CH_3)_3Sn{-}C{\equiv}C{-}C_6H_5 + MgBrCl \quad (136)$$

In a 250-ml three-necked flask equipped as in Fig. 19 (without drain) is prepared a solution of ethylmagnesium bromide from 10.9 g (0.10 mole) of ethyl bromide, 2.5 g (0.10 g-atom) of magnesium turnings, and 50 ml of anhydrous THF. The solution is treated with 10.2 g (0.10 mole) of freshly distilled phenylacetylene, and the mixture stirred for 16 hr at 25°. Then a solution of 19.9 g (0.1 mole) of chloro(trimethyl)tin in 50 ml of THF is added

[287] C. A. Kraus and W. H. Kahler, *J. Am. Chem. Soc.* **55,** 3537 (1933).

[288] J. J. Eisch, J. E. Galle, and M. Boleslawski, unpublished studies (1975).

over a 2-hr period. The mixture is heated at reflux for 16 hr, and then the solvent is removed *in vacuo*. The residue is extracted with three 75-ml portions of dry benzene, and the filtered extract freed of solvent and distilled. The trimethyl(phenylethynyl)tin distils at 72–74° at 0.5 mm Hg, 18.6 g (70%).

The nmr spectrum shows absorptions at (δ, CCl_4) 0.17 (s, 9H), 7.05 (m, 3H, meta and para H) and 7.15 (m, 2H, ortho H) ppm.

14. Tri-*n*-butyl(iodomethyl)tin

The introduction of functionalized methyl groups into a variety of metalloid derivatives is made immeasurably easier by the use of iodomethylzinc iodide.[289] By suitable treatment of a zinc–copper couple with methylene iodide this reagent can be isolated and subsequently coupled with the halides of germanium, tin, lead, and mercury.[290] In the case of tin, such iodomethyltins can be converted into a number of functionalized methyltins by simple nucleophilic displacements, and these latter derivatives in turn into functionalized methyllithium derivatives by tin–lithium exchange:

$$R_3Sn{-}CH_2{-}I \xrightarrow[-I]{:N} R_3Sn{-}CH_2{-}N \xrightarrow[-R_4Sn]{RLi} N{-}CH_2{-}Li \qquad (137)$$

Procedure[290b]*:*

$$CH_2I_2 \xrightarrow[THF]{Zn(Cu)} ICH_2ZnI \xrightarrow[-ZnClI]{(n\text{-}C_4H_9)_3SnCl} (n\text{-}C_4H_9)_3SnCH_2{-}I \qquad (138)$$

In a 300-ml three-necked flask (Fig. 19) are placed 187 mg of copper(II) acetate and 18.7 ml of glacial acetic acid. The stirred solution is warmed to 100° and treated with 12.2 g (0.19 g-atom) of granulated zinc. After 2 min the hot acetic acid is drained off and a fresh 18.7-ml portion of acetic acid added. After another 2-min period of shaking the hot suspension, the acid is removed and the cooled reddish-brown solid washed with three 50-ml portions of dry ether.

The zinc–copper couple is dried under a current of dry nitrogen and suspended in 33 ml of anhydrous THF. Then a small portion of a solution of 15.2 ml (50.5 g, 0.19 mole) of diiodomethane in 33 ml of THF is added to the stirred suspension, whereupon an exotherm and purple coloration show the start of reaction. Then 66 ml of THF is added to the reaction mixture, and the balance of the diiodomethane solution added at a rate adequate

---

[289] G. Wittig and F. Wingler, *Justus Liebigs Ann. Chem.* **656,** 18 (1962).

[290] (a) D. Seyferth and S. B. Andrews, *J. Organometal. Chem.* **30,** 151 (1971); (b) This procedure for tri-*n*-butyl(iodomethyl)tin was developed by J. E. Galle of our laboratory. Subsequent to our work the preparation of this compound was briefly described in the literature (W. C. Still, *J. Am. Chem. Soc.* **100,** 1481 (1978)).

to maintain the reaction mixture at 35–40°. (Gentle heating is applied toward the end of the 2-hr addition period.)

The iodomethylzinc iodide solution is filtered directly into a nitrogen-flushed 250-ml three-necked flask (Fig. 19, without drain) and then warmed to 35–40°. A solution of 27.0 ml (32.6 g, 0.10 mole) of tri-*n*-butyl(chloro)tin in 40 ml of THF is added dropwise to the warm zinc reagent, and the resulting mixture is heated at reflux for 11 hr. The reaction mixture is diluted with 150 ml of benzene and then extracted with five 100-ml portions of 5% aqueous hydrogen chloride. The organic layer is dried over anhydrous sodium sulfate, freed of solvent, and distilled. The tri-*n*-butyl(iodomethyl)tin is collected as a colorless liquid, bp 105–110° at 0.1 mm Hg, 39.3 g (90%).

The nmr spectrum of this stannane shows a singlet at 1.86 (2H) and a multiplet at 0.8–1.7 (27H) ppm.

15. 1,1-Di-*n*-butyl-4,5-dihydrostannepin

The following hydrostannation of 1,5-hexadiyne illustrates one that is promoted by a source of free radicals (azobisisobutyronitrile) and that yields both cyclic and linear polymeric products. The isolation of the stannepin by distillation involves, in part, thermal depolymerization of the linear products.[285,291]

Procedure[291]*:*

$$HC{\equiv}CCH_2CH_2C{\equiv}CH + (n\text{-}C_4H_9)_2SnH_2 \longrightarrow c\text{-}C_6H_8Sn(n\text{-}C_4H_9)_2 + \left[-\underset{n\text{-}C_4H_9}{\overset{n\text{-}C_4H_9}{Sn}}-CH{=}CH(CH_2)_2CH{=}CH- \right]_x \quad (139)$$

A 500-ml three-necked flask equipped as in Fig. 19 (without drain) is charged with 28 g (0.356 mole) of redistilled 1,5-hexadiyne and 50 mg of azobisisobutyronitrile. Under nitrogen 83.5 g (0.356 mole) of freshly distilled di-*n*-butyltin dihydride is slowly added and the resulting solution heated at reflux for 8 hr. The benzene is evaporated, and the residue distilled (Fig. 29) under reduced pressure. Some gas evolution occurs, keeping the oil pump pressure at 10–15 mm Hg. The distillation bath is kept below 250°, as the

[291] This procedure was developed by R. J. Wilcsek and J. E. Galle, 1973–1974.

1,1-di-*n*-butyl-4,5-dihydrostannepin distils at 120–160°, 68.9 g (62%), as a colorless liquid. The pot residue is a greenish-gray, mealy solid.

The nmr spectrum of the dihydrostannepin shows absorptions at (δ, $CDCl_3$) 0.8–2.6 (m, 18H), 2.32 (t, 4H, $J = 2$ Hz), 5.94 (d, 2H, $J = 13$ Hz), and 6.70 (broad d of d, $J = 13$ Hz) ppm.

16. Chloro(triphenyl)tin and Dichloro(diphenyl)tin

Organotin(IV) halides can be prepared by: (*a*) the interaction of tin with active alkyl halides, such as those of methyl, allyl, and benzyl[292]; (*b*) the partial alkylation of stannic chloride by aluminum alkyls[293]; (*c*) the halodealkylation of tetraorganotins by hydrogen halides, halogen, or inorganic halides[294]; and (*d*) the redistribution reaction between appropriate ratios of tetraorganotins and stannic halides[295]:

$$2C_6H_5CH_2Cl + Sn \xrightarrow[\Delta]{C_6H_5CH_3} (C_6H_5CH_2)_2SnCl_2 \quad (140)$$

$$(CH_3CH_2)_3Al + SnCl_4 \xrightarrow{Et_2O} (CH_3CH_2)_3SnCl + AlCl_3 \quad (141)$$

$$(CH_3)_3SnCH_2CH_3 + I_2 \longrightarrow (CH_3)_2(CH_3CH_2)SnI + CH_3I \quad (142)$$

$$R_4Sn + SnCl_4 \xrightarrow{180^\circ} 2R_2SnCl_2 \quad (143)$$

The most generally applicable way is to synthesize the tetraorganotin and make use of its redistribution with SnX_4.

Procedure:

Caution: Although organotin halides are not sensitive to oxidation, they do hydrolyze rather easily and their vapors are irritating to the skin. Hence all operations should be performed under anhydrous conditions, and all transfers should be done in an efficient hood.

$$3(C_6H_5)_4Sn + SnCl_4 \longrightarrow 4(C_6H_5)_3SnCl \quad (144)$$

In a 2-liter three-necked flask equipped with a stirrer, an addition funnel, and an air condenser surmounted by a calcium chloride drying tube are placed 127 g (0.30 mole) of dried, powdered tetraphenyltin (mp 225–227°) and 16.7 ml (37.3 g, 0.14 mole) of anhydrous stannic chloride. The contents are stirred and heated in an electric mantle to 225–230° (with a precalibrated pyrometer connected to the mantle) for 3 hr. The clear liquid deposits a pale-yellow solid upon cooling. Then 400 ml of dry petroleum ether, bp 77–115°, and some animal charcoal are added and the mixture stirred under

---

[292] K. Sisido and Y. Takeda, *J. Org. Chem.* **26,** 2301 (1961).
[293] W. P. Neumann, *Justus Liebigs Ann. Chem.* **653,** 157 (1962).
[294] W. J. Pope and S. J. Peachey, *Proc. Chem. Soc.* **16,** 42, 116 (1902).
[295] K. A. Kocheshkov, *Ber.* **62,** 996 (1929).

reflux. Filtration of the hot solution and chilling to 0° deposit 112 g (73%) of colorless rhombic needles, mp 99–103°. Recrystallization gives 102 g (66%) of pure chloro(triphenyl)tin, mp 104–105°.

Procedure:

$$(C_6H_5)_4Sn + SnCl_4 \longrightarrow 2(C_6H_5)_2SnCl_2 \quad (145)$$

This reaction is performed in the same manner described in the preceding procedure, except that 85.4 g (0.20 mole) of tetraphenyltin and 52.1 g (0.20 mole) of stannic chloride are stirred and heated in a 1-liter three-necked flask for 2 hr at 170–180°. The cooled mixture is extracted with 250 ml of hot petroleum ether, bp 30–40°, to which some animal charcoal is added. Filtration of the hot solution and chilling overnight deposit 89 g (65%) of product, mp 38–41°. Recrystallization gives 72 g (52%) of dichloro(diphenyl)tin, mp 41–42°.

17. Triphenyltin Hydride

This frequently used stannane can be prepared in a manner similar to di-*n*-butyltin dihydride (Section F,12). Care should be exercised to minimize its photochemical or oxidative decomposition.[296] It is often used for hydrostannation (Section F,20).

Procedure[296,297]*:*

$$2(C_6H_5)_3SnCl + LiAlH_4 \longrightarrow 2(C_6H_5)_3SnH + LiCl + AlH_2Cl \quad (146)$$

In a 1-liter three-necked flask equipped as in Fig. 19 (without drain) are placed 300 ml of anhydrous ethyl ether and 3.2 g (84 mmoles) of powdered, white lithium aluminum hydride. The slurry is cooled in an ice bath, and 78 g (200 mmoles) of powdered, dry chloro(triphenyl)tin is slowly introduced (the addition funnel is replaced by a rubber tube–flask arrangement) over 30 min. The bath is removed, and the mixture is stirred at room temperature for 6 hr. Then with cooling the slurry is cautiously and slowly treated with 200 ml of water (*Caution: hydrogen evolution*). The ether layer is separated, washed with aqueous ammonium chloride solution, and dried over anhydrous sodium sulfate. The hydride and its solutions are protected from light. The ether is evaporated *in vacuo*, and the residual hydride is promptly distilled under a nitrogen atmosphere at the oil pump. Triphenyltin hydride distils at 164–165° at 0.3 mm Hg to yield 52 g (75%) of colorless liquid.

The nmr spectrum of neat triphenyltin hydride shows the Sn—H singlet

---

[296] H. Gilman and J. Eisch, *J. Org. Chem.* **20,** 763 (1955).

[297] (a) G. Wittig, F. J. Meyer, and G. Lange, *Justus Liebigs Ann. Chem.* **571,** 167 (1951); (b) E. Amberger, H. P. Fritz, C. G. Kreiter, and M. R. Kula, *Chem. Ber.* **96,** 3270 (1963).

at (δ) 6.83, the meta and para proton multiplet at 6.92, and the ortho proton multiplet at 7.37 ppm.[297b]

18. Tri-*n*-butyl(phenoxymethyl)tin

This derivative is typical of the functionalized methyltins derivable from iodomethyltins by nucleophilic displacement at the methylene carbon (1). Since tin is also prone to nucleophilic attack (2), experimental conditions are crucial to isolating the desired methyltin in high yield:

$$R_3SnN + ICH_2^- \xleftarrow[(2)]{:N^-} R_3Sn{-}CH_2{-}I \xrightarrow[(1)]{:N^-} R_3Sn{-}CH_2{-}N + I^- \qquad (147)$$

Procedure[298]:

$$(n\text{-}C_4H_9)_3Sn{-}CH_2{-}I + C_6H_5OH + K_2CO_3 \longrightarrow (n\text{-}C_4H_9)_3Sn{-}CH_2{-}O{-}C_6H_5 + KHCO_3 + H_2O \qquad (148)$$

In a 500-ml three-necked flask (Fig. 19, in air) a mixture of 100 ml of dimethyl sulfoxide, 34.2 g (80 mmoles) of tri-*n*-butyl(iodomethyl)tin, 8.3 g (88 mmoles) of phenol, and 12 g (87 mmoles) of anhydrous potassium carbonate is stirred at room temperature for 48 hr and at 70–75° for 20 hr. The reaction mixture is diluted with 150 ml of ether–pentane and extracted with five 50-ml portions of water. The organic layer is dried over anhydrous magnesium sulfate, evaporated, and chromatographed on a silica gel column. Elution with hexane gives 23 g (72%) of colorless tri-*n*-butyl(phenoxy)tin, which is pure by nmr spectral analysis (δ, CCl_4): 0.75–2.65 (m, 27H), 4.1 (s, 2H), and 6.7–7.3 (m, 5H) ppm.

19. Allyl(triphenyl)tin and Diallyl(diphenyl)tin

Although symmetric tetraorganotins can be prepared from stannic halides and organomagnesium,[299] -lithium,[299] or -aluminum[300] reagents, unsymmetric organotins, $R_nSnR'_{4-n}$, are best prepared from Grignard reagents and the appropriate R_nSnCl_{4-n}.[301] Under the conditions of transalkylation, lithium and aluminum alkyls often produce R_4Sn in addition to the desired product. For the preparation of allylic or benzylic stannanes,

---

[298] This procedure was developed by J. E. Galle, 1978.

[299] R. K. Ingham, S. D. Rosenberg, and H. Gilman, *Chem. Rev.* **60,** 459 (1960).

[300] W. P. Neumann, *Angew Chem.* **75,** 225 (1963).

[301] H. C. Clark and R. C. Poller, *Can. J. Chem.* **48,** 2670 (1970).

direct reaction of the organic halide, magnesium, and tin chloride (Barbier method) often succeeds. But the cleaner reaction, higher yields, and greater versatility attained with preformed Grignard solutions favor the two-step procedure.[296]

Procedure[296]*:*

$$(C_6H_5)_3SnCl + CH_2{=}CHCH_2MgBr \longrightarrow (C_6H_5)_3SnCH_2CH{=}CH_2 + MgBrCl \qquad (149)$$

Caution: Many allylic and benzylic halides are lachrymators and skin irritants. Disposable gloves should be worn, and transfers carried out in an efficient hood.

In a 1-liter three-necked flask equipped as in Fig. 19 are placed 48.9 g (2.02 g-atoms) of magnesium turnings and 100 ml of dry ethyl ether. After initiation, the balance of a solution of 40.8 g of freshly distilled allyl bromide in 190 ml of ether is added dropwise to the vigorously stirred solution over a 2-hr period. After the solution has cooled, it is drained into a calibrated addition funnel previously flushed with nitrogen. Acid titration indicates a yield of 80–85%.

Then 0.10 mole of the Grignard solution in 190 ml of ether is added to a 1-liter three-necked flask (Fig. 19, without drain) under nitrogen, and the funnel replaced by a wide, thin-rubber tube connected to an Erlenmeyer flask containing 27.0 g (70 mmoles) of ground, dried chloro(triphenyl)tin. Via this tube the stannane is introduced, in portions, into the stirred Grignard solution over a 30-min period. After the vigorous reaction has subsided, the mixture is stirred for 2 hr and then treated with 150 ml of saturated aqueous ammonium chloride solution. Separation, drying over anhydrous sodium sulfate, and evaporation of the ether layer give the crude product, which is extracted with 100 ml of refluxing petroleum ether, bp 77–115°. The cooled extract yields 24 g (88%) of allyl(triphenyl)tin, mp 73–75°.

The nmr spectrum of allyl(triphenyl)tin shows vinyl absorptions at (δ, CCl_4) 4.74 (terminal H cis to C_2H), 4.89 (terminal H trans to C_2H), and 6.00 (C_2H) ppm.

Procedure[296]*:*

$$(C_6H_5)_2SnCl_2 + 2CH_2{=}CHCH_2MgBr \longrightarrow (C_6H_5)_2Sn(CH_2CH{=}CH_2)_2 + 2MgBrCl \qquad (150)$$

In a similar manner, 0.10 mole of the Grignard reagent is treated dropwise from the addition funnel with a solution of 17.2 g (50 mmoles) of dichloro(diphenyl)tin in 50 ml of dry ether. After workup in the foregoing manner, the residual product is distilled to give 13 g (71%) of diallyl(diphenyl)tin, bp 173–174° at 6 mm Hg, n_D^{20} 1.6025.

20. Triphenyl(*trans*-β-styryl)tin

Hydrostannation is an important, broadly applicable method for synthesizing unsymmetric stannanes from olefins or acetylenes.[302–304] The addition of tin–hydrogen bonds to unsaturated systems can proceed spontaneously but in many cases requires promotion by a transition-metal or free-radical agent. With terminal unsaturated substrates, the addition is highly regioselective, giving the terminally bonded tin product. Stereoselectivity is also often attained, with the trans- or anti adduct predominating.

Procedure[305]:

$$(C_6H_5)_3Sn{-}H + C_6H_5{-}C{\equiv}C{-}H \longrightarrow \begin{matrix} C_6H_5 \\ H \end{matrix}\!\!>\!C{=}C\!<\!\!\begin{matrix} H \\ Sn(C_6H_5)_3 \end{matrix} \quad (151)$$

Caution: Tin hydrides are sensitive to oxygen, heat, and light, hence they do not store well. A freshly prepared or purified sample of triphenyltin hydride should be employed.

In a 100-ml Schlenk vessel (Fig. 13c) flushed with nitrogen and provided with a septum and magnetic stirring is placed 20 g (57 mmoles) of pure triphenyltin hydride. Cautiously, a few drops of 5.8 g (57 mmoles) of redistilled phenylacetylene are added via the septum, and the mixture stirred until heat is evolved (do not add the balance until reaction begins, for the reaction becomes extremely vigorous). The rest of the phenylacetylene is added in small portions as the exotherm subsides. Upon standing, the mixture crystallizes, and it is then recrystallized from 250 ml of hexane to yield 18.3 g (71%) of triphenyl(trans-β-styryl)tin, mp 117–119°. The nmr spectrum displays vinyl absorptions at (δ, CCl_4) 6.43 (d, $J = 13$ Hz) and 7.80 (d) ppm.

21. 1,1-Dimethyl-2,3,4,5-tetraphenylstannole

As mentioned in Section B,8, crucial to the successful synthesis of many metalloles is the proper preparation of the dilithium precursor. This approach has been used to prepare metallacyclopentadienes containing As, Au, B, Ge, Hg, Sb, Se, Si, Sn, Te, Tl, and Zn centers.[42,43] However, it should be noted that the structures claimed for some of the isolated products are either unproved (Hg)[43] or clearly in error (B).[155–156]

---

[302] H. Schumann and I. Schumann, *in* "Gmelin Handbuch der Anorganischen Chemie" (H. Bitterer, ed.), Organotin Compounds, Vol. 35/4, Springer Verlag, Berlin and New York, 1976.

[303] G. J. M. van der Kerk and J. G. Noltes, *Ann. N.Y. Acad. Sci.* **125,** 25 (1965).

[304] W. P. Neumann, "The Organic Chemistry of Tin," pp. 85–117, Wiley (Interscience), New York, 1970.

[305] G. J. M. van der Kirk and J. C. Noltes, *J. Appl. Chem.* **9,** 106 (1959).

Procedure[155–156]:

$$2C_6H_5{-}C{\equiv}C{-}C_6H_5 + 2Li \xrightarrow{Et_2O}$$

$$\text{(1,2,3,4-tetraphenyl-1,4-dilithio-1,3-butadiene)} \xrightarrow[-2LiCl]{(CH_3)_2SnCl_2} \text{(1,1-dimethyl-2,3,4,5-tetraphenylstannole)} \qquad (152)$$

In a 1-liter three-necked flask equipped as in Fig. 19 (except with magnetic stirring) are placed 53.4 g (0.30 mole) of diphenylacetylene and 250 ml of anhydrous ethyl ether. Then, under a coating of mineral oil, 2.09 g (0.30 g-atom) of lithium wire is cut into pieces, hammered out into a thin foil, and added to the flask. After 2 hr of stirring at room temperature the remaining lithium pieces are removed (1.62 g or 0.23 g-atom of lithium consumed). Then the red solution is diluted with 500 ml of anhydrous and deoxygenated THF to yield a green solution. This solution is stirred for 15 min.

Meanwhile, a 1.5-liter two-necked flask equipped with a large pressure-equalized addition funnel and a nitrogen inlet is charged with 25.3 g (0.115 mole) of dichloro(dimethyl)tin and 100 ml of anhydrous THF (in general, the amount of dichloro(dimethyl)tin used should be one-half the molar amount of lithium metal *consumed*). The lithium reagent is transferred, in portions, to the addition funnel and added rapidly to the stirred stannane solution, which is cooled in an ice bath. If any dark color of the lithium reagent persists after admixing the solution, small additions of dichloro-(dimethyl)tin can be made (usually 0.5–1.0 g) to discharge this color. The solvent is evaporated *in vacuo*, and the solid residue washed at *room temperature* with small amounts of 95% ethanol to remove any lithium salts or residual tin halides.[306]

The extracted residue is recrystallized from a methylene chloride–95% ethanol pair to yield colorless needles of 1,1-dimethyl-2,3,4,5-tetraphenylstannole, mp 185–188°. The yield is 76% if based on the acetylene, or 99% if based on the lithium consumed.

The nmr spectrum of this stannole shows absorptions at (δ, $CDCl_3$) 0.60 (s, 6H) and 6.7–7.2 (m, 20H) ppm.

---

[306] This washing at room temperature is important; otherwise, at higher temperatures, such salts will solvolyze with ethanol and produce hydrochloric acid, which will cleave the stannole to yield chloro(dimethyl)1,2,3,4-tetraphenyl-1,3-butadien-1-yltin.

In a test experiment, brief heating of 2 mmoles each of the stannole and dichloro-dimethyl)tin in a 1:1 mixture of methylene chloride and 95% ethanol gave an 85% yield of the butadienyltin. For this reason, it is important to use an amount of dichloro(trimethyl)-tin determined by the lithium consumed and to remove residual amounts from the crude product by an ethanol extraction *without heating*.

22. Halo(triphenyl)lead and Dihalo(diphenyl)lead

The preparation of organolead halides is best achieved by the controlled cleavage of tetraorganoleads by halogen or hydrogen halide. Concentration of reagents and contact time are important in favoring the mono- over the dihalo product:

$$R_4Pb \xrightarrow[-RE]{E-X} R_3PbX \xrightarrow[-RE]{E-X} R_2PbX_2, \qquad E = H, X \tag{153}$$

Grüttner and Krause recommend the use of bromine in pyridine solution at $-50°$ as a means of preparing bromo(triphenyl)lead in 90% yield.[307,308] The drawback of such a solvent and the convenience of employing hydrogen halide to make either mono- or dihalide recommend the methods of Gilman described here.[309,310]

Procedure[309,310]*:*

$$(C_6H_5)_4Pb \xrightarrow[-C_6H_6]{HCl} (C_6H_5)_3PbCl \xrightarrow[-C_6H_6]{HCl} (C_6H_5)_2PbCl_2 \tag{154}$$

Caution: All organolead compounds should be viewed as toxic. Disposable gloves should be worn to protect the skin from absorbing any solution. However, a nitrogen atmosphere is not required.

In a 1.5-liter three-necked flask equipped as in Fig. 19 (without drain, but with an 8-mm gas inlet tube in place of the addition funnel) are placed 50.5 g (98 mmoles) of tetraphenyllead and 700 ml of dry chloroform. The contents are stirred under reflux to form a solution, and then dry hydrogen chloride gas is led into the gently boiling solution for ~30 min or until a white precipitate [dichloro(diphenyl)lead] begins to form. The introduction of hydrogen chloride is immediately stopped, and reflux continued for ~30 min in order to consume any dissolved hydrogen chloride. The solution is filtered to remove the dichloride (1–2 g), and the chloroform evaporated from the filtrate. The residue is extracted with two 500-ml portions of absolute ethanol [residue of starting $(C_6H_5)_4Pb$], and the filtered, cooled extracts yield 32.5 g of chloro(triphenyl)lead, glistening needles, mp 204–205°. Concentration of the filtrate gives another 2–3 g of product for a total yield of 73–75%.

Procedure[309,310]*:*

In a similar apparatus, 51.5 g (0.1 mole) of tetraphenyllead dissolved in 500 ml of benzene at 50–60° is treated with gaseous hydrogen chloride over

[307] G. Grüttner, *Ber.* **51,** 1298 (1918).
[308] E. Krause and O. Schlöttig, *Ber.* **58,** 429 (1925).
[309] H. Gilman and J. D. Robinson, *J. Am. Chem. Soc.* **51,** 3112 (1929).
[310] H. Gilman and D. S. Melstrom, *J. Am. Chem. Soc.* **72,** 2953 (1950).

a 60-min period. To ensure that all the dichloride has been formed, the mixture is filtered *hot* (to keep the monochloride and tetraphenyllead in solution) and the hot filtrate retreated with hydrogen chloride gas until no more precipitate is formed. The combined dichloro(diphenyl)lead amounts to 38–43 g (89–99%).

The following procedure exemplifies how both mono- and dibromides can be isolated from such a cleavage reaction. By varying the amount of hydrogen bromide gas introduced, the ratio of products can be changed.[311]

Procedure:

$$(C_6H_5)_4Pb \xrightarrow[-C_6H_6]{HBr} (C_6H_5)_3PbBr \xrightarrow[-C_6H_6]{HBr} (C_6H_5)_2PbBr_2 \qquad (155)$$

A 2-liter three-necked flask equipped as in the foregoing procedures is charged with 40 g (78 mmoles) of tetraphenyllead and 750 ml of chloroform. The mixture is stirred and heated to incipient reflux while a rapid stream of dry hydrogen bromide gas is passed into the solution. After 5–10 min a white precipitate of dibromo(diphenyl)lead is formed. The solution is cooled, and 25 g (62%) of dibromide is filtered off (decomposes above 225°). The chloroform filtrate is evaporated to dryness, and the residue is recrystallized from absolute ethanol to yield 10.5 g (26%) of bromo(triphenyl)lead as rhombic needles, mp 165–167°.

23. Allyl(triphenyl)lead

Tetraorganoleads, especially those bearing unsaturated groups, are thermally unstable in many cases.[296] Tetravinyllead, for example, is reported to explode above 110°.[312] In a less dramatic manifestation of instability, unsymmetric organolead compounds tend to disproportionate upon heating or in the presence of Lewis acids:

$$R_nPbR'_{4-n} \xrightarrow{\Delta} R_4Pb + R'_4Pb \qquad (156)$$

For these reasons, the preparation and purification of these organometallics should avoid high temperatures (>100–150°), and such Lewis acids as LiX, MgX_2, and AlX_3 should be removed before warming any crude product. Despite such measures, the inherent instability of certain derivatives cannot be surmounted. Thus, although both allyl- and benzyl(triphenyl)lead are stable, readily crystallized derivatives, diallyl- and dibenzyl(diphenyl)lead are too unstable to be isolated pure, even at temperatures below 40°. The crude diallyl(diphenyl)lead decomposes readily over 100° to produce lead metal, tetraphenyllead, allyl(triphenyl)lead, and biallyl.[296]

---

[311] J. J. Eisch, unpublished studies.

[312] E. C. Juenge and S. E. Cook, *J. Am. Chem. Soc.* **81,** 3578 (1959).

Procedure[296,313]:

$$(C_6H_5)_3PbCl + CH_2{=}CHCH_2MgBr \longrightarrow (C_6H_5)_3PbCH_2CH{=}CH_2 + MgBrCl \quad (157)$$

In a 500-ml three-necked flask equipped as in Fig. 19 (except that the additional funnel is replaced by a wide rubber tube connected to an Erlenmeyer flask) is placed 45 ml of 1.14 *M* allylmagnesium bromide in ether (51 mmoles). Then, with stirring, 11.3 g (24 mmoles) of powdered chloro(triphenyl)lead is added in portions, via the rubber tube, over 30 min. The mixture turns from white to dark gray over 60 min. After stirring overnight the mixture is hydrolyzed with aqueous ammonium chloride solution, and the separated organic layer dried over anhydrous magnesium sulfate. Removal of the volatiles and recrystallization of the residue from 95% ethanol gives colorless needles of allyl(triphenyl)lead, mp 75–77°, 9.6 g (84%).

24. Tetraphenyllead

Symmetric tetraorganoleads cannot feasibly be made like other Group IVA derivatives because lead(IV) salts are too unstable. Instead, lead(II) salts are treated with organolithium, -magnesium, or -aluminum reagents and the resulting R_2Pb disproportionated into R_4Pb and finely divided lead. By using the corresponding organic iodide, the deposition of lead is avoided in organolithium reactions[314,315]:

$$4RMgI + 2PbCl_2 \longrightarrow R_4Pb + Pb + 2MgCl_2 + 2MgI_2 \quad (158)$$
$$3RLi + PbCl_2 + RI \longrightarrow R_4Pb + 2LiCl + LiI$$

In addition, various commercial processes involving either the action of lead alloys on alkyl halides or the electrolysis of organometallic reagents at a lead anode have been developed for the production of tetramethyl- and tetraethyllead. The use of these compounds as gasoline antiknock additives has been decreasing in recent years because of the introduction of automobile emission controls.[316,317]

Procedure[314]:

$$3C_6H_5Br + 6Li \longrightarrow 3C_6H_5Li + 3LiBr \quad (159)$$
$$3C_6H_5Li + PbCl_2 + C_6H_5I \longrightarrow (C_6H_5)_4Pb + 2LiCl + LiI$$

Caution: Contact with vapors or solutions of organolead compounds should be avoided, since they are highly toxic. Although vapors are not a

[313] P. R. Austin, *J. Am. Chem. Soc.* **53,** 3514 (1931).
[314] H. Gilman and R. G. Jones, *J. Am. Chem. Soc.* **72,** 1760 (1950).
[315] R. W. Leeper, L. Summers, and H. Gilman, *Chem. Rev.* **54,** 101 (1954).
[316] W. J. Considine, *Ann. N.Y. Acad. Sci.* **125,** 4 (1965).
[317] L. C. Willemsens, "Organolead Chemistry," International Lead-Zinc Research, New York, 1964.

problem with the nonvolatile compounds used here, disposable gloves should be worn and all vessels rinsed carefully right after use.

A 1.5-liter three-necked flask equipped as shown in Fig. 19 is flushed with dry nitrogen and charged with 200 ml of anhydrous ethyl ether and 12.0 g (1.73 g-atoms) of freshly cut lithium pieces (2–3 mm, cut and dropped directly into the flask). About 40 drops of a solution of 126 g (0.8 mole) of bromobenzene in 400 ml of ether is added to initiate the reaction (turbidity, exotherm, shiny spots on the lithium). The balance of the bromobenzene is added over 2 hr, while maintaining the stirred mixture at gentle reflux by periodic cooling (using a bath and a cloth collar containing ice around upper part of the flask). The solution is stirred for another 30 min and filtered under nitrogen into a calibrated, nitrogen-flushed addition funnel. Analysis of a hydrolyzed aliquot for total alkali shows a 97% yield of phenyllithium.

A 2-liter three-necked flask equipped as in Fig. 19 (without drain) is charged with 69.5 g (0.25 ml) of dried, powdered lead(II) chloride, 56.0 (0.275 mole) of iodobenzene, and 200 ml of anhydrous ethyl ether. Under nitrogen, the foregoing ethereal solution of phenyllithium (0.77 mole) is added to the stirred suspension over 25 min with no external cooling. After 5 min the mixture turns bright yellow; then the color slowly turns to brownish gray. The mixture is stirred for 12 hr at 20–25° and then for 3 hr at reflux. The cooled mixture is slowly poured into ice water, and the gray solid collected and dried. The crude product is extracted for 30 hr in a Soxhlet extractor with 600 ml of chloroform. The chloroform deposits 72 g (56%) of colorless, glistening needles of tetraphenyllead, mp 228–229.5°. By evaporating the chloroform mother liquor, combining with the Soxhlet residue, and re-extracting the combined solid with 250 ml of chloroform, another 31 g of product of the same melting point is obtained. The total yield is 80%.

Appendix I. Bibliography of Compilations on Organometallic Chemistry

A. Textbooks

1. G. E. Coates, M. L. H. Green, P. Powell, and K. Wade, "Principles of Organometallic Chemistry," Methuen, London, 1968.
2. G. E. Coates and K. Wade, "Organometallic Compounds," 3rd ed., Vol. I, Main Group Metals; Vol. II (M. L. H. Green), Transition Metals, Methuen, London, 1968.
3. J. J. Eisch, "The Chemistry of Organometallic Compounds" (main groups emphasized), Macmillan, New York, 1967.
4. R. B. King, "Transition Metal Organometallic Chemistry—An Introduction," Academic Press, New York, 1970.
5. P. L. Pauson, "Organometallic Chemistry" (emphasis on transition metals), St. Martin's Press, New York, 1967.
6. E. G. Rochow, "Organometallic Chemistry" (105 page paperback), Van Nostrand-Reinhold, Princeton, New Jersey, 1964.
7. E. G. Rochow, D. T. Hurd, and R. N. Lewis, "The Chemistry of Organometallic Compounds," Wiley, New York, 1957.

B. Monographs

1. B. J. Aylett (ed.), "MTP International Review of Science. Inorganic Chemistry," Vol. 4, Organometallic Derivatives of the Main Group Elements, University Park Press, Baltimore, Maryland, 1972.
2. J. N. Friend (ed.), "Textbook of Inorganic Chemistry" (A. E. Goddard and D. Goddard, 1928); Vol. XI. Organometallic Compounds. Part I: Groups I–IV. Part II: Arsenic (A. E. Goddard, 1930); Part III: Group V–VIII (A. E. Goddard, 1936), Griffin, London.
3. E. Krause and A. von Grosse, "Die Chemie der Metallorganischen Verbindungen," Bornträger, Berlin, 1937.
4. A. N. Nesmeyanov and K. A. Kocheshkov (eds.), "Methods of Elemento-Organic Chemistry" (English translation by North Holland Publ., Amsterdam). Vol. I: B, Al, Ga, In, Tl (A. N. Nesmeyanov and R. A. Sokolik, 1967); Vol. II: Be, Mg, Ca, Sr, Ba (A. N. Nesmeyanov and S. Ioffe, 1967); Vol. III: Zn, Cd (N. I. Sheverdina and K. A. Kocheshkov, 1967); Vol. IV: Hg (L. G. Markarova and A. N. Nesmeyanov, 1968).
5. H. H. Zeiss (ed.), "Organometallic Chemistry" (a collection of reviews), Van Nostrand-Reinhold, Princeton, New Jersey, 1960.

C. Review Series

1. E. W. Abel and F. G. A. Stone (eds.), "Chemical Society Specialist Reports, Organometallic Chemistry" (annual survey, 1971–1980, Vol. 1–8). The Chemical Society, London.
2. E. I. Becker and M. Tsutsui (eds.), "Organometallic Reactions," Vol. 1–5, Wiley (Interscience), New York, 1970–1975.
3. D. Seyferth and R. B. King (eds.), "Organometallic Chemistry Reviews" (journal), Elsevier, Amsterdam, 1963–1971. Merged into *J. Organometal. Chem.* in 1972. Part A, Subject Reviews; Part B, Annual Surveys (henceforth separate issues of *J. Organometal. Chem.* are devoted to annual surveys).
4. F. G. A. Stone and R. West (eds.), *Adv. Organometal. Chem.* **1–16,** (1964–1977).
5. Other continuing sources of reviews: (1) *Accounts of Chemical Research;* (2) *Advances in Inorganic Chemistry and Radiochemistry;* (3) *Angewandte Chemie;* (4) *Chemical Reviews;* (5) *Chemical Society Reviews;* (6) *Coordination Chemistry Reviews;* (7) *Fortschritte der chemischen Forschung* (*Topics in Current Chemistry*); (8) *Inorganica Chimica Acta Reviews;* (9) *Inorganic Syntheses;* (10) *Organic Reactions;* (11) *Progress in Inorganic Chemistry;* (12) *Pure and Applied Chemistry* (publication of principal lectures presented at international symposia); (13) *Quarterly Reviews;* (14) *Russian Chemical Reviews;* and (15) *Transition Metal Chemistry.*

D. Compendia of Compounds

1. N. Hagihara, M. Kumada, and R. Okawara (eds.), "Handbook of Organometallic Compounds," Benjamin, New York, 1968.
2. M. Dub (ed.), "Organometallic Compounds," 2nd ed., Springer Verlag, 1966, Vol. I, Transition Metals; Vol. II, Ge, Sn, Pb; Vol. III, As, Sb, Bi (literature coverage 1937–1964).
3. M. Becke-Goehring and K. C. Buschbeck (eds.), "Gmelin Handbuch der Anorganischen Chemie," Erganzungswerk zur 8. Auflage (New Supplement Series), Springer Verlag, Berlin and New York. The following volumes are devoted to organometallic compounds and are written mostly in German; some sections and all major headings are given in English; Ag (Part B, Section 5, 1975); B (carboranes, Vol. 15, 27, 42 änd 43, 1974–1977), borazines, Vol. 51, 1978); Bi (Vol. 47, 1977); C (graphite metal compounds, Part B, Section 3, 1968); Co (Vol. 5 and 6, 1973); Cr (Vol. 3, 1971); F (perfluorohaloorganic metal derivatives, Vol. 24 and 25, 1975); Fe (Vol. 14, 36, 41, 49 and 50, 1974–1978); Hf (Vol. 11, 1973); Ni (Vol. 16–18, 1975); Sn (Vol. 26, 29, 30 and 35, 1975–1976); Si (Part C, 1958); Ti (Vol. 40, 1977); V (Vol. 2, 1971); and Zr (Vol. 10, 1973).
4. V. Bazant, V. Chvalovsky, and J. Rathousky (eds.), "Organosilicon Compounds" (three volumes; literature coverage through 1961), Publishing House, Czechoslovak Academy of Sciences, 1965.

E. Specialized Books on Main-Group Metals

1. Group I

(a) D. J. Cram, "Fundamentals of Carbanion Chemistry," Academic Press, New York, 1965.
(b) D. S. Matteson, "Organometallic Reaction Mechanisms of the Nontransition Elements," Academic Press, New York, 1974.
(c) O. A. Reutov and I. P. Beletskaya, "Reaction Mechanisms of Organometallic Compounds," Wiley (Interscience), New York, 1968.

2. Group IIA

M. S. Kharasch and O. Reinmuth, "Grignard Reactions of Nonmetallic Substances," Prentice Hall, Englewood Cliffs, New Jersey, 1954.

3. *Group IIB*
F. R. Jensen and B. Rickborn, "Electrophilic Substitution of Organomercurials," McGraw-Hill, New York, 1968.
4. *Group III*
Boron
(a) H. C. Brown, "Hydroboration," Benjamin, New York, 1962.
(b) H. C. Brown, "Boranes in Organic Chemistry," Cornell Univ. Press, Ithaca, New York, 1972.
(c) H. C. Brown, "Organic Syntheses Via Boranes," Wiley (Interscience), New York, 1975.
(d) W. Gerrard, "The Organic Chemistry of Boron," Academic Press, New York, 1961.
(e) R. Grassberger, "Organische Borverbindungen," Verlag Chemie, 1971 (paperback).
(f) R. N. Grimes, "Carboranes," Academic Press, New York, 1970.
(g) E. L. Muetterties (ed.), "The Chemistry of Boron and Its Compounds," Wiley, New York, 1967.
(h) E. L. Muetterties and W. H. Knoth, "Polyhedral Boranes," Dekker, New York, 1968.
(i) T. Onak, "Organoborane Chemistry," Academic Press, New York, 1975.
(j) H. Steinberg, "Organoboron Chemistry," Vol. I, B—O, B—S (1964); Vol. II, B—N; Vol. III, B—C. Pergamon Press, Oxford.
(k) H. Steinberg and A. L. L. McCloskey (eds.), "Progress in Boron Chemistry," Pergamon, Oxford, 1964.
Aluminum
T. Mole and E. A. Jeffrey, "Organoaluminium Compounds," Elsevier, Amsterdam, 1972.
5. *Group IV*
(a) E. Y. Lukevits and M. G. Voronkov, "Organic Insertion Reactions of the Group IV Elements" (English translation), Consultants Bureau, New York, 1966.
(b) A. G. MacDiarmid (ed.), "Organometallic Compounds of the Group IV Elements," Vol. 1, M—C bonds (1968); Part 1: Si—C, Part 2, Ge—C, Sn – C, Pb—C; Vol. 2, M—X bonds (1972); Part 1. Si—X, Part 2. Ge—X, Sn—X, Pb—X, Dekker, New York.
Silicon
(a) C. Eaborn, "Organosilicon Compounds," Buttersworths, London, 1960.
(b) A. D. Petrov, V. F. Mironov, and E. A. Chernyshev, "Synthesis of Organosilicon Monomers" (English translation), Consultants Bureau, New York, 1964.
(c) E. G. Rochow, "Chemistry of the Silicones," 2nd ed., Wiley, New York, 1951.
(d) L. H. Sommer, "Stereochemistry, Mechanism and Silicon," McGraw-Hill, New York, 1965.
Germanium
(a) F. Glockling, "The Chemistry of Germanium," Academic Press, New York, 1969.
(b) M. Lesbre, P. Mazerolles, and J. Satge, "The Organic Compounds of Germanium," Wiley, New York, 1971.
Tin
(a) W. P. Neumann, "The Organic Chemistry of Tin," Wiley, New York, 1970.
(b) R. C. Poller, "The Chemistry of Organotin Compounds," Academic Press, New York, 1970.
(c) A. K. Sawyer (ed.), "Organotin Compounds" (3 Vols.), Dekker, New York, 1971–1972.
Lead
H. Shapiro and R. W. Frey, "The Organic Compounds of Lead," Wiley, New York, 1968.
6. *Group V*
G. O. Doak and L. D. Freedman, "Organometallic Compounds of Arsenic, Antimony and Bismuth," Wiley (Interscience), New York, 1970.

7. *Group VI*
D. L. Klayman and W. H. H. Gunther (eds.), "Organic Selenium Compounds—Their Chemistry and Biology," Wiley (Interscience), New York, 1973.

F. Synthesis and Experimental Technique

1. C. J. Barton, Glove box techniques, *in* "Techniques of Inorganic Chemistry" (H. B. Jonassen and A. Weissberger, eds.), Vol. III, p. 259, Wiley (Interscience), New York, 1963.
2. T. R. Crompton, "Chemical Analysis of Organometallic Compounds," Vol. 1–5, Academic Press, New York, 1973–1977.
3. J. J. Eisch and R. B. King (eds.), "Organometallic Syntheses"; Vol. 1, R. B. King, Transition-Metal Compounds," 1965; Vol. 2, J. J. Eisch, "Nontransition-Metal Compounds," 1981; Academic Press, New York.
4. S. Herzog, J. Dehnert, and K. Luhder, An expedient method for preparative procedures in an inert atmosphere, *in* "Technique of Inorganic Chemistry" (H. B. Jonassen and A. Weissberger, eds.), Vol. VII, p. 119, Wiley (Interscience), New York, 1968.
5. Houben-Weyl's "Methoden der Organischen Chemie," Vol. XIII, Organometallic Compounds, Georg Thieme Verlag, Part 1: Groups IA and IB, Part 2a: B, Mg, Ca, Sr, Ba, Zn, Cd; Part 2b: Hg; Part 4: Al, Ga, In, Tl; Part 7: Pb, Ti, Zn, Hf, V, Nb, Ta, Cr, Mo, W; Part 8: As, Sb, Bi, 1970–1978.
6. "Inorganic Syntheses," Vol. 1–18, Wiley (Interscience), 1939–1978.
7. G. W. Kramer, A. B. Levy, and M. M. Midland, Laboratory Operations with Air-Sensitive Substances: Survey, *in* "Organic Synthesis via Boranes" (H. C. Brown, ed.), Wiley (Interscience), New York, 1975.
8. H. Metzer and E. Müller, Arbeiten unter Ausschluss von Sauerstoff und Luftfeuchtigkeit, *in* "Methoden der Organischen Chemie" (Houben-Weyls, ed.), Vol. I/2, Georg Thieme Verlag, 1959.
9. "Organic Syntheses," Vols. 1–57 and Coll. Vols. 1–5, Wiley, New York, 1940–1978.
10. D. F. Shriver, "The Manipulation of Air-Sensitive Compounds," McGraw-Hill, New York, 1969.

G. Safety and Hazard

1. J. M. Barnes and L. Magos, *Organometal. Chem. Rev.* **3,** 137 (1968).
2. L. Bretherick (ed.), "Handbook of Reactive Chemical Hazards," 2nd ed., Butterworths, London, 1978.
3. H. E. Christenson and T. T. Luginbyhl, The Toxic Substance List, U.S. Dept. Health, Education, and Welfare, Washington, D.C., 1974.
4. L. Friberg and J. Vostal, "Mercury in the Environment," Chemical Rubber Company, Cleveland, Ohio, 1972.
5. N. I. Sax, "Dangerous Properties of Industrial Materials," Van Nostrand-Reinhold, Princeton, New Jersey, 1975.
6. "Toxic and Hazardous Industrial Chemicals Safety Manual," International Technical Information Institute, Tokyo, Japan, 1975.
7. K. Verschueren, "Handbook of Environmental Data on Organic Chemicals," Van Nostrand-Reinhold, Princeton, New Jersey, 1977.
8. R. C. Weast (ed.), "CRC Handbook of Laboratory Safety," 2nd ed., Chemical Rubber Company, Cleveland, Ohio, 1971.

Appendix II. Suppliers of Reagents for Organometallic Chemistry

Given below is a listing of chemical companies that offer organometallics, metals, reagents, and solvents suitable for organometallic synthesis. Where particular types of organometallics or reagents are a company's specialty, it is indicated in parentheses. To learn which companies sell a specific organometallic, one should consult the most recent edition of "Chemical Sources—USA," issued annually by Directories Publishing Company, Inc., Flemington, N.J. 08822. However, in searching for a given organometallic in this compilation, one must check the companies listed under each of several possible names for the compound. Many of the companies listed here do custom syntheses of chemicals not listed in their current catalogs. Current mailing addresses may be found in "Chemical Sources—USA."

Abbott Laboratories, Chemical Division, D-902 (Hg)
Aceto Chemical Company (Fe)
Adams Chemical Company (Cr)
Aeroceuticals (Hg)
Aldrich Chemical Company, Inc. (organometallics used in organic synthesis; many organometallics)
Amerchol, A Unit of CPC International, Inc. (Mg)
American Cyanamid Company (Co, Fe)
American Cyanamid Company, Fine Chemicals Dept. (Co)
American Drug and Chemical Company (Mg, Hg)
American Research Products Company (Co)
American Roland Corporation (Co)
Amersham/Searle (As)
Analabs, Inc., Subsidiary of New England Nuclear
The Ansul Company (As)
Apache Chemicals, Inc. (Al, B)
Arapahoe Chemicals, Inc., A Syntex Company (Mg)
Atomergic Chemetals Corporation (many organometallics, Ga)
Beecham Lab (Hg)
Biochemical Laboratories, Inc. (As)
Bio Clinical Laboratories, Inc. (As, Hg)
Bodman Chemicals (many organometallics)
Briar (Hg)
Brinkman Instruments (Hg)
Calbiochem (Co)
Callery Chemical Co., Division of Mine Safety Appliance Co. (B, alkali metals)
Century (Hg)
Chembond Corporation (Al, Li)
Chemical Dynamics Corporation (Hg)
Chemical Procurement Labs, Inc. (many organometallics)

Chemical Samples Company (Mg)
Chem Service, Inc. (As)
Chicago Pharmaceutical (Hg)
City Chemical Corporation (many organometallics)
Columbia Organic Chemical Company, Inc. (organic halides, Ge)
Davis and Sly (Hg)
Degussa, Inc. (cyano derivatives)
Delamar, Inc. (As)
D. F. Goldsmith Chemical and Metal Corporation (Ir, Hg, Os)
Diamond Shamrock Corporation (As)
Dormar Chemicals, Inc. (Co)
Dow Corning Corporation (Si)
Eastern Chemical, Division of Guardian Chemical Corporation (As)
Eastman Organic Chemicals, Eastman Kodak Company (many organometallics)
Electronic Space Products (Al)
Electron Microscopy Sciences (As)
EM Laboratories, Inc., Affiliate of E. Merck, Darmstadt, Germany (many organometallics)
Englehard Minerals and Chemicals Corporation, Englehard Industries Division (transition metals)
Ethyl Corporation, Industrial Chemicals Division (Al)
Fairfield Chemical Company (As, Mg)
Fallek Chemical Corporation (Co)
Filo Color and Chemical Corporation (Fe)
Fisher Scientific Company (many organometallics)
Foote Mineral Company, Chemical and Minerals Division (Li)
Freeman Industries, Inc. (Hg)
Ganes Chemicals, Inc. (Hg)
General Electric Company, Silicone Products Dept. (Si)
George Uhe Company, Inc. (Bi)
G. Frederick Smith Chemical Company (nitrogen ligands)
Grain Processing Corporation (Co)
Guardian (Hg)
Hardwicke Chemical Company, Subsidiary of McLaughlin Gormley King Company (Mg)
Haven Chemicals, Division of Haven Industries, Inc., Special Organic Chemicals (Fe, Li)
Houston Chemical Company (Pb)
Hynson, Westcott and Dunning (Hg)
I.C.N./K & K, Life Sciences Group (many organometallics)
I.C.N. Pharmaceuticals, Life Sciences Group (As, Bi)
J. T. Baker Chemical Company (As)
King's Laboratory, Inc. (Hg)
Lachat Chemicals, Inc. (As)
Lakewide (Hg)
Laramie Chemical Company (Ge, Mg)
Lemmon (Hg)
Lilly (Hg)
Lithium Corporation of America (Li)
Mallinckrodt, Inc. (Hg)
M & T Chemicals, Inc. (Sn)
Marine Colloids, Inc. (As)
Marshallton Research Laboratories, Inc. (Si)
MC & B Manufacturing Chemists

Mead-Johnson (Hg)
Merck and Company, Inc., Merck Chemical Division (Co, As, Hg)
Mine Safety Appliances (alkali metals)
MWM Chemical Corporation (Co)
Ohio Valley Specialty Chemical Company (Si)
Orgment, Inc. (many main group, phosphines, some transition)
Orion Chemical Company (many organometallics)
Parish Chemical Company (Fe)
Parke, Davis and Company, Subsidiary of Warner Lambert Company (As)
PCR, Inc. (many organometallics, Si)
Petrarch Systems (Si)
Pfaltz and Bauer, Inc., Division of Aceto Chemical Company (many organometallics)
Pfizer, Inc. (Co)
Pharm-Eco Laboratories, Inc. (Hg)
Pierce Chemical Company (Hg)
Polychemical Laboratories (As, Bi)
Polysciences, Inc. (many organometallics)
Pressure Chemical Company (carbonyls, arsines, phosphines)
Raylo Chemicals, Ltd. (B)
Regis Chemical Company (Si)
Reliable Chemical Company (many organometallics)
Research Organic/Inorganic Chemical (many organometallics)
Research Plus Laboratories, Inc. (As, Co)
Rhodia Inc., Chemicals Division, Chemical Dept. (As, Hg)
Roussel Corporation (Co, Hg)
RSA Corporation (As)
Ruger Chemical Company (Hg)
Schwarz/Mann, Division of Becton, Dickson and Company (Hg)
Scientific Gas Products, Inc. (Ni)
Sigman Chemical Company (As, B, Hg)
Silar Laboratories, Inc. (Si)
Simmler and Son, Inc. (Hg)
Standex (Hg)
Stauffer Chemical Company, Specialty Chemical Division (Al)
Sterling Organics (As)
Strem Chemicals, Inc. (transition, carbonyls, arsines, phosphines)
Synthathron Corporation (Ga)
Texas Alkyls, Inc. (Al, Mg)
Trans World Chemicals, Inc. (Fe)
Tridom Chemical, Inc. (many organometallics)
Troy Scientific Corporation (Hg)
Union Carbide Corporation, Chemicals and Plastics (Si)
United States Biochemical Corporation (As, Hg)
Ventron Corporation, Chemicals Division, Alfa Products (many organometallics)
Vineland Chemical Company, Inc. (As, Mg)
VWR Scientific (many organometallics)
W. A. Cleary Corporation (As, Hg)
Wateree Chemical Company, Inc. (Hg)
Whitehall (Hg)
Whitmoyer Laboratories, Inc. (As)
Wyeth (Hg)

Appendix III. Special Solvents for Organometallic Reagents*

Solvent	Formula	Boiling point (°C)	Melting point (°C)	Remarks
Hydrocarbons				
Cyclopentane	C_5H_{10}	49	−94	Low-temperature nmr
Cyclohexane	C_6H_{12}	81	6.6	Cryoscopic solvent
Methylcyclohexane	C_7H_{14}	111	−127	Low-temperature nmr
Mesitylene	C_9H_{12}	165	−45	High temperatures for Group III R_nM
Tetralin	$C_{10}H_{12}$	208	−36	High temperatures for Group III R_nM
Ethers				
Dimethyl ether	$(CH_3)_2O$	−24	−39	Medium for alkali metal–hydrocarbon adducts[a]
1,4-Dioxane	$O(CH_2CH_2)_2O$	102	12	Preparation of halide-free R_2Mg or R_3In[b,c]
Dimethoxymethane	$H_2C(OCH_3)_2$	42	−105	Preparation of labile RLi reagents[d]
2-Methyltetrahydrofuran	$(CH_2)_3CHCH_3O$	80	—	Substitute for THF[e]
Tetrahydropyran	$(CH_2)_5O$	88	−45	Substitute for THF[e]
Phenetole	$C_6H_5OCH_2CH_3$	172	−30	Medium for air-oxidation of RMgX[f]
Triethylene glycol–dimethyl ether	$CH_3O(CH_2CH_2O)_3CH_3$	222	−45	High-temperature solvent substitute for diglyme
Di-*n*-butyl ether	$(CH_3CH_2CH_2CH_2)_2O$	142	−98	Medium for Zerwitinoff analysis[g]
Diphenyl ether	$(C_6H_5)_2O$	258	27	High-temperature medium of low coordinating tendency[h]
Crown ethers	$(O{-}C_n{-})_n$	—	—	Coordination of Group IA ions[i]

* Addendum to Table I.

Solvent	Formula	Boiling point (°C)	Melting point (°C)	Remarks
Nitrogen donors				
Ammonia	NH_3	−33	−78	< −33°, medium for Group IVA reactions
Acetonitrile	CH_3CN	82	−46	†
Dimethylformamide	$HCON(CH_3)_2$	152	−61	†
N,N-Dimethylacetamide	$CH_3CON(CH_3)_2$	165	−20	†
Tetramethylurea	$(CH_3)_2NCON(CH_3)_2$	167	−1	†
Triethylamine	$(CH_3CH_2)_3N$	89	−115	†
1,4-Diazabicyclooctane	$N(CH_2CH_2)_3N$	—	160	
Quinoline	C_9H_7N	238	−16	†
Isoquinoline	C_9H_7N	243	27	†
Halides				
Carbon tetrachloride	CCl_4	77	−23	‡
1,2-Dichloroethane	$ClCH_2CH_2Cl$	84	−35	‡
Chlorobenzene	C_6H_5Cl	132	−46	‡
o-Dichlorobenzene	$C_6H_4Cl_2$	181	−17	‡
Special				
Dimethyl sulfide	$(CH_3)_2S$	37	−98	Solvent for organomagnesium compounds[j]
Diethylzinc	$(CH_3CH_2)_2Zn$	118	−28	Complexing solvent for Group IA organometallics[k]

[a] N. D. Scott, J. F. Walker, and V. I. Hansley, *J. Am. Chem. Soc.* **58,** 2442 (1936).

[b] W. Schlenk and W. Schlenk, Jr., *Ber.* **62B,** 920 (1929).

[c] I. M. Viktorova, Y. P. Endovin, N. I. Sheverdina, and K. A. Kocheshkov, *Dokl. Akad. Nauk SSSR* **177,** 103 (1967).

[d] W. Tochtermann and H. Küppers, *Angew. Chem.* **77,** 173 (1965).

[e] H. Normant, *Bull. Soc. Chim. Fr.* 1434 (1963).

[f] H. Gilman and A. Wood, *J. Am. Chem. Soc.* **48,** 806 (1926).

[g] W. Fuchs, N. H. Ishler, and A. G. Sandhoff, *Ind. Eng. Chem. Anal. Ed.* **12,** 507 (1940).

[h] J. J. Eisch and W. C. Kaska, *J. Am. Chem. Soc.* **88,** 2976 (1966).

[i] C. J. Pedersen, *J. Am. Chem. Soc.* **89,** 7017 (1967); **92,** 386, 391 (1970).

[j] G. Bähr and K. H. Thiele, *Chem. Ber.* **90,** 1578 (1957).

[k] F. Hein, E. Petzchner, K. Wagler, and F. A. Seitz, *Z. Anorg. Allgem. Chem.* **141,** 161 (1924).

† Saturated nitrogen solvents are suitable for RLi and organometallics of Groups IIA–IVA; solvents containing C=O, C=N, or C≡N will decompose R_nM of Groups IA and IIA; all these solvents coordinate strongly with R_nM of Groups IA–IIIA (except Hg, B) and alter the properties of the organometallic.

‡ Suitable for organometallics of Group IVA but generally should not be used with R_nM of Groups I–III (*danger of explosion*).

Appendix IV. Periodic Table of the Main-Group Elements

PERIODIC TABLE OF THE MAIN–GROUP ELEMENTS

Groups

I	II	IIB	III	IV	V	VI	VII	O
1 H 1.0079								2 He 4.003
3 Li 6.94	4 Be 9.012		5 B 10.81	6 C 12.011	7 N 14.007	8 O 15.9994	9 F 18.998	10 Ne 20.179
11 Na 22 99	12 Mg 24.305		13 Al 26.982	14 Si 28.086	15 P 30.974	16 S 32.06	17 Cl 35.453	18 Ar 39.948
19 K 39.098	20 Ca 40.08	30 Zn 65.38	31 Ga 69.72	32 Ge 72.59	33 As 74.922	34 Se 78.96	35 Br 79.904	36 Kr 83.80
37 Rb 85.468	38 Sr 87.62	48 Cd 112.41	49 In 114.82	50 Sn 118.69	51 Sb 121.75	52 Te 127.60	53 I 126.91	54 Xe 131.30
55 Cs 132.91	56 Ba 137.33	80 Hg 200 59	81 Tl 204.37	82 Pb 207.2	83 Bi 208.98	84 Po (209)	85 At (210)	86 Rn (222)
87 Fr (223)	88 Ra 226.03	The atomic weights are based upon the relative atomic mass of $^{12}C=12$						

Index of Compounds